THE ROOTS OF LIFE

A Layman's Guide to Genes, Evolution, and the Ways of Cells

Mahlon B. Hoagland

Houghton Mifflin Company Boston 1978

Library of Congress Cataloging in Publication Data

Hoagland, Mahlon B
The roots of life.

Includes index.
1. Biology. I. Title.
QH309.H63 1978 574 77-13722
ISBN 0-395-25811-1

Printed in the United States of America

V 10 9 8 7 6 5 4 3 2 1

For Olley

Foreword

It has been my great good fortune to have participated actively in the recent exciting discoveries of modern biology, leading to our present understanding of the laws governing life processes. The simplicity and beauty of the truths that have come to light are a source of pleasure and wonderment to me. In the course of teaching these laws to medical students and explaining them to my nonscientist friends, I have become convinced that the enjoyment of the experience can be shared by most people — even those who are unschooled in science. In this matter, my wife, to whom this book is dedicated, has been a passionate advocate. Indeed, her insistence that scientists have an obligation to open their work to the uninitiated has had a strong influence in motivating and shaping this little book.

Up until the present century scientists generally worked as independent entrepreneurs, supported by benevolent monarchs or wealthy patrons, occasionally by their government, sometimes by their own modest means. The rising expectations of citizens everywhere, and the heightened complexity and cost of scientific technology, have brought scientists into a much closer partnership with government and people. As knowledge grows and becomes increasingly pertinent and applicable to human practical concerns, people want to know what scientists are doing: Are they accomplishing what is expected of them and, for that matter, are they doing things they should not be doing? It would be easy, for example, to say that because science is perceived to be failing us, or frightening us, or not fulfilling all of our hopes, government support should be withdrawn. There may be instances where, after careful study, this is deemed proper. More often, however, such action would be like taking lights away from explorers in the depths of an unknown cave — remembering, of course, that we're all groping in the cave together! The wise management of such problems urgently needs a fuller understanding by all citizens and their legislative representatives of what medical science is, what it can and can't do.

These general considerations have had a part in my writing this book. Another reason has to do with the challenge implicit in the fact that most nonscientists are mired in the conviction that science is incomprehensible; by its very nature, abstruse and mysterious. Although the unwillingness of most professional scientists to adapt their vocabulary to the layman's needs is a factor here, of more importance is the intellectual languor of the citizen not schooled in science. But surely this aversion to science is an unnatural state of mind. Science is the natural searching within ourselves and in our surroundings for explanations. It is the process of making comprehensible, by discovering and explaining with simple laws, that which had been dark — and often frightening — mystery. It is curiosity in harness, and curiosity is surely one of the most elemental human motivating forces. As children, we practice a kind of science by freely and unabashedly exercising our curiosity. As we age, we suppress the very drive that had given us much enjoyment and knowledge. I welcome the challenge and pleasure of making science pleasurable for others.

My final reason for telling this story relates to my role as president and scientific director of a basic biomedical research institute, The Worcester Foundation for Experimental Biology. My days are full of déjà vu as my younger colleagues reawaken in me the excitement, wonder, and sense of progress that is basic research. This book celebrates both the achievements of the past and the commitment of a new generation to advance the boundaries of our knowledge as one continuum. And the foundation's many generous friends, who have worked so devotedly with me to create a superb environment for continuing our work, are certainly deserving of this tangible expression of my appreciation of their valued moral and material support.

I am grateful to Paul C. Zamecnik and Bernard D. Davis who helped me to find the wonder and excitement of molecular biology; to my wife, Olley, whose unwavering faith in the value of this enterprise made it happen; to Jacqueline Foss for cheerfully and skillfully typing repeated drafts of the manuscript; to Debra Grega whose beautiful free-hand lettering adorns my crude illustrations; and to David and Ann Barlow, Clarence Berger, Polly Cowan, Winifred Chrisman, Judith Hauck, Hudson Hoagland, Jennifer and Bill Mason, Judith Pederson, and Violetta Scornik for their help in improving the manuscript.

Vinalhaven
1977

Contents

The Roots of Life

CHAPTER I

Simplicity Is the Sign of Truth

I remember long ago a walk with my father on a lonely beach. The sea was gray; ragged clouds scudded before a chill, early winter wind. It was a day for discovery. Lying in the decaying seaweed at the high-tide border were washed-up, empty bottles of all shapes and sizes. And it gradually dawned on us that every bottle was capped — we couldn't find a single topless bottle! We were puzzled by such unwonted conformity among bottles until my father struck upon the explanation. He delightedly urged me to see in the bottles a grander meaning. The upshot was a lesson in evolution, firmly screwed into my consciousness for life. For obviously those bottles were the few survivors of an ocean journey, the fit few. Of the many empty bottles thrown into the sea by the hand of man, a very few would have had a top replaced by some inadvertent, chance act, rendering them unsinkable. The vastly larger number of nonsurvivors, caplessly ill-suited to the ocean's hostility, would have long since sunk to the bottom.

The creative act in science is a little like that inspiration of the bottles of my youth. It is the envisioning of a simple law or explanation that shepherds a welter of meaningless and bewildering facts into order. So it was with Charles Darwin and Alfred Russel Wallace over a century ago. Separately pondering, they had sought to explain the, until then, bewilderingly wide distribution and diversity of living things on the earth. Their magnificently simple concept of the survival of those best suited to their particular environment by virtue of fortuitous changes in the way they were constructed was a brilliant leap of the imagination. Their inspired explanation rendered simple and comprehensible a vast accumulation of biological information.

But inspired insight is useless without meticulous observation. And meticulous observation is pedantry if the searcher doesn't know how to generate ideas about the observations and then test the ideas by experiment. If the ideas are good and the experiments clever, the answers are likely to reveal something hitherto unknown.

The broad general questions asked by religionists and philosophers lead to broad general conclusions, seldom verifiable and therefore not universally acceptable. Direct, concise questions put to nature tend to give unequivocal, simple answers, answers that can be confirmed by others. Scientific knowledge is thus built from the bottom up, small pieces at a time, each piece a firm support because it is *true*, that is, universally verifiable. I am defining *truth* in terms acceptable to the scientist. It will be argued that human experience reveals other kinds of truth. But such truths are not expected to meet the standards of confirmation by others, verifiability. And so, although they are truths to some, they are opinions to others.

This book will focus on the simple principles that govern the state and process of being alive. These laws of life illuminate all of biology and medicine. They make it easier for us to comprehend the meaning of being alive in all its seemingly unsimple ramifications. And they are aesthetically pleasing. *Simplex sigillum veri*, the Romans said. "Simplicity is the sign of truth."

Focus on the Cell

If we are to study the things about life that are simple and true, we must begin with the cell. For the cell is the minimum organizational structure of life in all its forms. *There is nothing alive that is simpler than a cell, and nothing can start to get more complex without first being a cell.*

Let me try to justify this important statement by dissecting you, the reader, part by part until I get to the essence of your aliveness — that irreducible component of your make-up that is common to *all* living creatures.

1. You are *conscious* of yourself, your surroundings, what you are doing, including the fact that you're reading this book. Consciousness is an activity of a very special organ, the brain, and is about as complicated a biological phenomenon as exists. We know practically nothing about how consciousness works and probably won't for a long time, if ever, but we suspect that the great majority of living things experience no consciousness of their brains.

2. Your body is a community of organs — brain, heart, lungs, liver, kidneys, muscles, bones, skin, endocrine glands — all operating *together* in precisely regulated harmony. Your body temperature normally never varies from 98.6°F (37°C). Your brain, nerves, and muscles coordinate your movements and keep you in perfect balance; you economically take in food and oxygen and

eliminate waste, keeping your total weight constant. Most of these superlative coordinating abilities are shared with almost all other mammals, as well as with birds, frogs, and fishes. On the other hand, many of the simpler forms of life manage quite well without them.

3. Each of your individual organs or tissues is a living part of you, made up of large populations of cells, each doing its special thing. Your brain cells become long and threadlike, and completely devote themselves to conducting electrical messages from one place to another; skin cells become tough and elastic to act as your body's outer protection; bone cells accumulate calcium phosphate within them to make them rigid so that they can support your body; and so on. Like worker bees and soldier ants, specialized cells are committed to doing a limited number of tasks in service to the larger whole, you. The all-important job of reproducing yourself in your children becomes the assignment of a special collection of your cells. Cell specialization is very widespread among living creatures, including plants and many simpler forms of life in the sea. But many tiny creatures don't get together with other cells and specialize; they retain remarkable versatility and remain independently capable of self-reproduction and of satisfying their nutritional needs from very simple substances.

4. Not so long ago your own cells were autonomous, substantially more versatile and freely self-duplicating — like the simplest forms of life. This was when they had just become settled in the wall of your mother's uterus. They were not recognizably you, but all the information to create you was there. Every single cell in that tiny mass of cells descendent from the fertilized egg contained the prescription for you and had already inaugurated the construction project that would eventually be you.

It is the way you were when you began in your mother that brings into sharp focus your kinship with all other creatures. Your start in life has special meaning for our explorations in this book. It affirms that *every living thing, from highest to lowest, was or is a cell capable of dividing to make more copies of itself.* More complex, "higher," many-celled creatures must make single cells (eggs and sperms) when they want to make copies of themselves. The big difference among cells from different creatures is their content of "information," which will instruct their internal machinery to become a bacterium, a mosquito, a frog, or a human.

What Cells Are Made Of

So our attention is to be focused upon the cell as the simplest, minimum arrangement of things that has the quality of aliveness. Let's be sure, then, that we have a clear idea of the things a cell is made of. Here they are in order of increasing complexity.

1. *Atoms.* The five principal atoms we should know about are carbon, hydrogen, oxygen, nitrogen, and phosphorus. There are many others in much smaller amounts. Atoms are the natural elements of the universe and are the smallest entities from which life is constructed. The five principal atoms of life have an average weight of 15 atomic weight units. So we'll call them size 15, for comparison with the larger molecules, which follow.

2. *Simple Molecules.* These are combinations of atoms. Molecules in living cells are sometimes called organic molecules. There are hundreds of different kinds of molecules in cells. On the average they are about size 150, or ten times bigger than atoms.

3. *Chain Molecules.* Still called molecules, these are simple molecules connected so as to make long chains. The most important of these chains average about size 75,000: 500 times bigger than simple molecules; 5000 times bigger than atoms. The largest of the linked molecules, which reach several million units, can be seen by the most powerful electron microscopes.

4. *Structures.* These are linked molecules fitted into architectural arrangements within the cell. The smallest structure is about size 7.5 million, or 100 times bigger than the average linked molecules. The larger structures are 10 or more times larger than that. Structures can be seen by ordinary light microscopes.

5. *Cells.* These are, as we've said, the minimum organizations of structures that are alive. Most cells are too small to be seen by the naked eye but can easily be seen in a weak ordinary microscope, or even a good magnifying glass.

6. *Organs.* These are groups of cells working cooperatively at a specialized task within an organism.

7. *Organisms.* An organism is the minimum arrangement of cells necessary for the complete functioning of any particular form of life. A bacterial cell or a yeast cell is an organism because a single cell is all that's necessary for these creatures to be autonomously themselves, especially to reproduce themselves. Humans are organisms that require the harmonious participation of some 60 trillion cells to be complete entities!

Order amid Chaos

Putting atoms together to make molecules, connecting molecules to make chains, arranging chains to make structures, and arranging structures to make living cells is a colossal organizational job — bigger by far than anything that humans can accomplish with their brains, hands, and computers. Yet this incredible job is being accomplished every moment all over the earth. Indeed, at the very root of life is the living cell's constant dedication to the job of creating and maintaining order, organization, complexity.

Physicists tell us that the inanimate universe is steadily becoming more disorganized; everything is moving slowly — on a time scale of billions of years — toward chaos. The second law of thermodynamics states that *entropy* — the official word for disorder — is steadily increasing everywhere in the universe.

Why should the universe "aspire" to disorganization? It isn't as strange as it first appears. Look at it this way. Suppose you had some dilute blue paint and some dilute yellow paint and you poured them into a single container. The paint molecules will bounce about, as is the way of molecules, until there is an even green mixture. The molecules have become completely randomly distributed, disorganized; have reached what is for them the most stable configuration. If you desired to push the system backward so as to return to an *un*even, nonrandom, organized mixture of blue and yellow molecules — say, the blue on top and the yellow on the bottom — you'd be working against the system's strong "desire" to reach the random, stable, disorganized state of greenness.

So it is with all atoms and molecules in the universe. They seek the Nirvana of randomness, of total disorder, of ultimate stability. The castle you build of sand is evanescent and is inevitably destined to become featurelessly flat. Volcanoes speak for the earth's tumultuous search for the stability of sameness. Rocks become imperceptibly sands;

and sands, the salts of the sea. Inexorably, all things move toward randomness, which is, for all the universe's atoms and molecules, final stability.

Now, while the states of randomness and of stability are identical for inanimate objects, we humans intuitively find it hard to see the two as equivalent. And that is quite natural, because life's whole thrust is against nature's drive to randomize. Life works constantly to create *un*stable states. Life works against randomness; life creates order. Life, on an infinitely grander scale, regularly does the equivalent of *de*mixing green paint.

Energy Is Needed to Make Order

A process working constantly against nature's drive to make disorder can be successful only with help — help in the form of energy. Construction of the elegant, wondrously complex interior of a tiny cell requires energy. The energy comes from the sun.

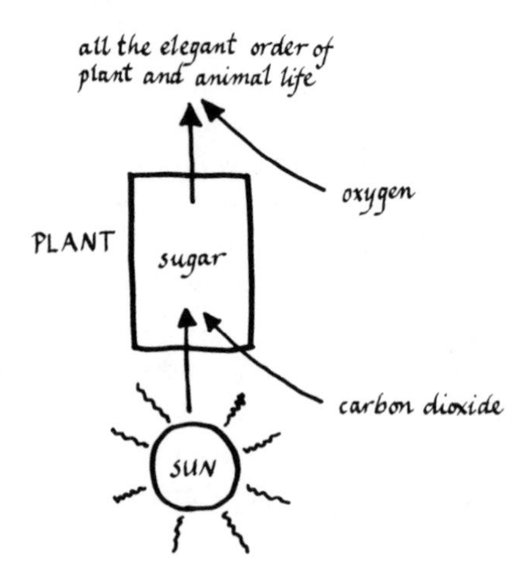

Plants trap sunlight. The trapped light is used to convert carbon dioxide to sugar. Sugar is more complex than carbon dioxide; that is, it is more ordered. So sunlight creates order by running the sugar-making machine. Sugar is the universal food for all living creatures. Since sugar's order required energy (sunlight) for the making, the breaking down of sugar will give energy back. Plants and animals break down sugar by "burning" it with oxygen, thus releasing carbon dioxide. The energy so produced is used by plants and animals to construct their own substance — all the complex structures of living cells.

Sugar, then, by virtue of its ordered structure, provides energy that is used to create much more living order. The order of a living cell is many many thousands of times greater than the order of a sugar molecule; so, to make the equation balance, live things must consume many, many thousands of sugar molecules to create themselves. And this they do.

Being alive, therefore, not only means order, organization, complexity; more important, it means the ability to create and maintain order and organization in a hostile environment that is working against it. This is one sense in which the creation of a new life has the quality of a miracle.

A Plan Is Needed to Make Order

Is life really the only thing in the universe that can actually generate order? Water, when it cools, becomes a solid, and the ice molecules can arrange themselves in elaborate and intricately beautiful ordered patterns. A salt in solution can crystallize out of solution, a process that increases the order among the salt molecules. There are many isolated examples, but they are all singularly unimpressive compared with the accomplishments of the simplest living cell. Furthermore, the creation of order by a living cell is fundamentally different from these crystallization processes. Cells make order by following a preexisting plan.

It is reasonable to suppose that before an arrangement of things in space can be made, there must be a plan. Something must have a "preconception" of the anticipated ordering of parts.

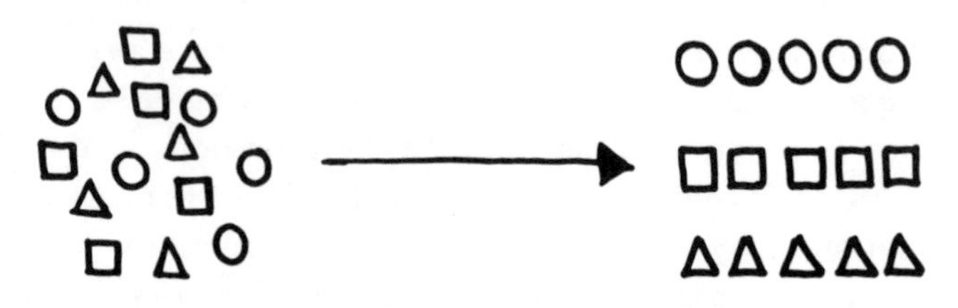

For example, if it is required that the squares, circles, and triangles pictured in the left panel be arranged in some order that is not just random, someone or something has to know what to do; must have conceived beforehand the arrangement on the right. Given the plan, the task theoretically can be accomplished. All that's needed is energy, as we've already determined.

What is the nature of the information that could instruct a cell how to arrange its atoms, molecules, chains, structures in the correct patterns so as to become a complete living cell, and that could be passed down the generations to keep the process going? Well, I find it awesome that we know the explanation. And it is marvelously clever and simple. The revelation of the nature of biological information is surely the most exciting insight of modern biology and one of the most important events in all of the history of science. We discuss it next.

CHAPTER II

Information

We gaze upon the faces of our children as into a mirror. Our children are our own reincarnations. We provide, in sperm and egg, detailed instructions about our own physical make-up, which our children's cells use to make themselves in our likeness. Those instructions are our gift to the future, a gift of information.

Information for making a cell must have the quality of a map, a plan, or a blueprint; or a brochure, manual, or book. It must be the complete instructions that can be "comprehended" by an agent or machine capable of performing the specified construction task: the making of a life.

Genes

The science of genetics has learned that each feature or trait (for example, eye color) of a living being is inherited; that is, reproduced accurately in the offspring. Information that prescribes individual traits resides in special entities called *genes.* For every distinct inheritable trait there is a separate gene. Gregor Mendel, the founder of the science of genetics, had, by 1860, shown that genes were inherited as though they were real things; that is to say, they couldn't be diluted, subdivided, or mixed during inheritance. Genes, then, seemed to be little inheritable packets of information, each governing a particular trait of an organism.

In the 1920s the great geneticist Thomas Hunt Morgan found where the genes were actually located in cells. Within all cells there is a container called the nucleus. The nucleus was known to be very important since it had to divide before a cell divided to become two. So the process of sharing the wealth of one cell equally among the offspring began in the nucleus. Furthermore, the microscope revealed within the nucleus threadlike structures called chromosomes. These structures doubled themselves before the nucleus divided, and one set of chromosomes went to each of the "daughter" cells. Because of this arrangement, it was suspected that the chromosomes were the location of genes. Morgan proved this was indeed the case in a series of elegant experiments using common fruit flies as experimental "animals." Before he completed his great work, it was known that genes were in fact strung along the chromosome threads like beads.

What Are Genes Made Of?

That was known by the 1930s. Soon scientists were asking the exciting question: What were chromosomes (genes) made of?

The experiments of a biologist named Oswald Avery, certainly among the most significant in biology, gave a brilliantly clear answer to that question. His work opened the modern era of what we now call molecular biology. In the early 1940s, Avery was concerned with the bacteria that caused "double" pneumonia — that major cause of death before the introduction of penicillin. He was seeking an explanation for the fascinating observation that dead pneumonia-causing bacteria could transmit their capacity to cause pneumonia to living, closely related non-pneumonia-causing bacteria. That is, dead dangerous bacteria could make living benign bacteria become dangerous. And once this transformation had occurred it was permanent, and was inherited by all future generations of the once-benign bacteria.

The capacity to cause disease is a genetic trait, or group of traits. These traits are controlled by, and inherited as, genes. Avery reasoned that the dead bacteria were disintegrating and their bodies were releasing chemical information-containing material that was being consumed by the living bacteria.

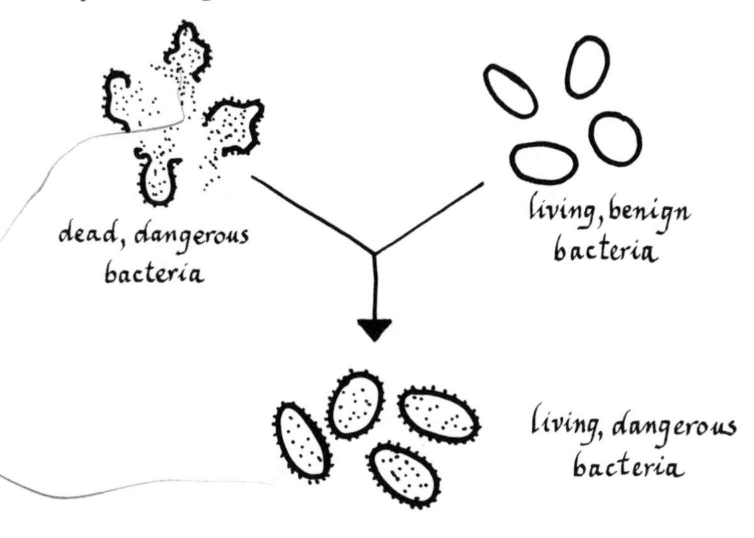

That is, genes were getting into the living bacteria and subsequently determining their inheritance. Avery and his fellow scientists set out to identify unequivocally this genelike material.

One could hardly conceive of a more important problem for medical science: discovering the chemical identity of genes. But it was certainly not a problem that could be studied in humans, and probably not even in animals. These pneumonia-causing bacteria presented Avery with an ideal system. They're an excellent example of the value of a good experimental model system. Indeed, the whole edifice of genetics, from its beginnings over 100 years ago with Gregor Mendel right up to current active research, was largely built on simple experimental models: peas, fruit flies, bread molds, and bacteria. All the cells Avery worked with were genetically identical; that is, pure-bred. They could be grown rapidly so that inheritance of traits could be followed over many generations in a short time and their ability to produce pneumonia could be measured with facility when they were injected into mice.

One of the most important experiments Avery did gave a very clear answer. He took the mixture of molecules released from the dead bacteria and added an enzyme that *destroys* DNA. The destruction of the DNA abolished the mixture's ability to transform benign bacteria to disease-causing bacteria. With additional experimentation, Avery and his colleagues proved that the material that transformed benign bacteria to disease-producing bacteria was *deoxyribonucleic acid*, or DNA.

DNA: Deoxyribonucleic Acid

Now, Avery hadn't discovered DNA. That had been done sixty years earlier by a researcher named Friedrich

Miescher. He and scientists to follow him had amassed much chemical information about it. It was known to be constructed of molecules called nucleotides, linked together in chains and containing large amounts of phosphoric acid. It was an enormous molecule, the largest ever found in cells. What Avery's discovery showed was that DNA was the actual material basis of heredity. That is, *inheriting something meant getting a piece of DNA.* Genes *are* DNA. Information is DNA and DNA is information.

From the time of Avery's proof, knowledge about DNA grew at such an astonishingly rapid pace that by 1960 we knew how information was encoded in DNA, how such information was converted into cell substance, and how DNA was copied so as to be shared with the next generation. There were many scientists involved in this incredible tour de force, but the inspired stroke of James Watson and Francis Crick in imagining the correct structure of DNA to be a double helix, two intertwined chains, was the biggest forward leap by far.

Here, then, are the key features of DNA:

1. It is a chain molecule (that is, a substance made up of different kinds of simple molecules connected together to make chains).

2. It is enormously long and extremely thin. If the cell's nucleus were to be magnified 100 times, it would be about the size of a pinpoint just visible to the eye. The DNA folded up in that nucleus would be the length of a football field!

3. There are four kinds of links (molecules called *nucleotides*) in the chain. Their names are adenylic acid, guanylic acid, cytidylic acid, and thymidylic acid; the abbreviations are A, G, C, and T.

4. The manner of connecting the four links is identical, as in any ordinary chain.

5. There is an exact order of the links, like the order of letters in a book.

From now on we'll be having a lot to say about chains. Each time we picture a chain we'll use one of five forms

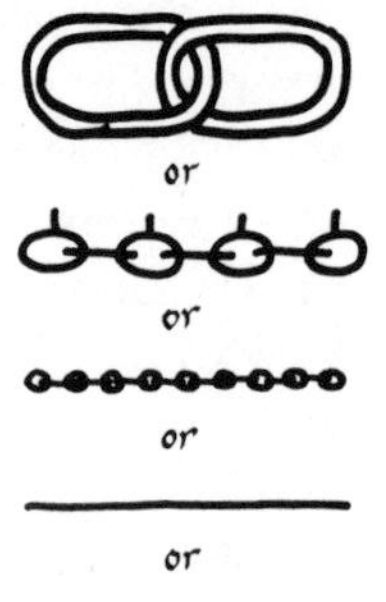

depending on which is most useful at the time. Of course the real chains are very much longer than our pictures suggest.

DNA = Language = Information

Now, if you have a chain made of links of four different kinds and you know it contains all the information for the construction of a new individual, you must conclude that

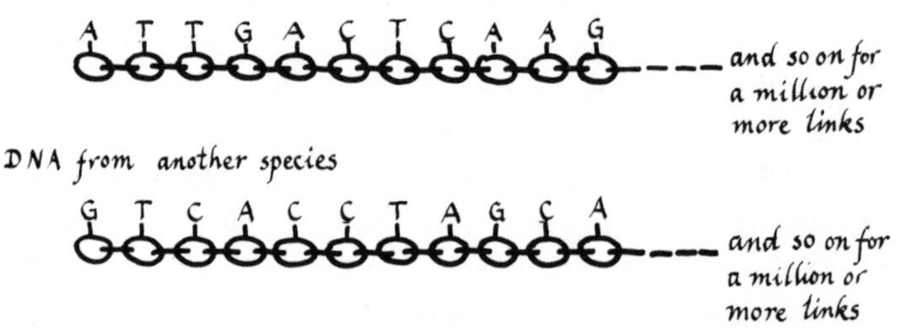

the secret lies in the sequence, or order, of the links. There's just no other way to explain how a chain can have that much meaning. Information turns out, then, not to be a map or a blueprint, a thing of two dimensions on a flat surface, but rather a set of "written" instructions in one dimension. The analogy with language is entirely apt. The DNA alphabet contains only four letters, but the number of messages that can be written with it is infinite — just as there is no limit to what can be said using the two-letter alphabet (dot and dash) of Morse code.

Letters in books are linked in sequence by their position on the paper; the four nucleotide links in DNA are connected in sequence by actual chemical bonds. The total amount of DNA in a given organism may be thought of as a book in which all the letters, words, phrases, sentences, and paragraphs are strung together to make one long chain that "means" all the parts and functions of that organism. That organism's identical twin contains identical DNA — an identical book; no letter or word differs. Another organism of the same species contains a similar book, though there are frequent and noticeable differences in grammar. The book of a different species tells a quite different story, though many sentences will still be similar.

In the above analogy, genes, which are sections of the chain, are approximately equivalent to sentences. A gene is a letter (nucleotide) sequence specifying a particular structure or function of an organism. Genes, like sentences, are strung out end to end in the very long DNA molecule.

How Much Information to Make a Human?

Having learned what information is, let's get a rough idea of how much information is needed to make living things.

1. A bacterium, one of the simplest of living creatures, has about 2000 genes; each gene has about 1000 letters (links) in it. So the bacterium's DNA must be at least 2 million letters in length.

2. A human being has over 500 times as many genes as a bacterium, so the DNA must be at least 1 billion letters in length.

3. The bacterium's DNA would be equivalent to 20 average novels, each of 100,000 words, and the human's to 10,000 such novels!

From Language to Substance

The meaning of the language of DNA lies in its specification of a particular living organism. In other words, the genes provide instructions for the creation of substance, the actual stuff of life, an actual living thing. How is DNA's language converted physically into living, breathing, moving, reproducing flesh? To answer this question we must first know what we're made of.

Proteins

This doesn't present as formidable a problem as it might seem. Far and away the most important material is protein. The rest of the things in us — water, salts, vitamins, metals, carbohydrates, fats, and so on — are there to help the proteins. Proteins are not only most of our bulk (not counting water), but they underlie our warmth, actions, thoughts, and feelings — indeed, everything we are and everything we do. I, for instance, observe my cat: her total bulk is protein. What I see mostly — fur, eyes, yes, even the motion — is protein. The stuff inside is protein. But, in addition, everything that makes my cat a very particular individual personality is determined by special proteins. Proteins made on the instructions of DNA are the physical basis of individuality, uniqueness, species. Protein to us is as metal is to an automobile. There are other materials in a car, but the dominant structural and functional element is metal. Metal determines both the appearance of the car and its ability to operate. And the shape, quality, and position of metal parts determine how one car differs from another.

Now, one more question and we're ready for a new insight. What are proteins made of? Here's a list of features:

1. They are chain molecules.
2. They are long, but not nearly as long as DNA.
3. There are *twenty* kinds of links (called amino acids).
4. The manner of connecting the twenty units is identical.
5. The order or sequence of the twenty units, or links, is exact, and determines what the protein is — what its ultimate special function will be.

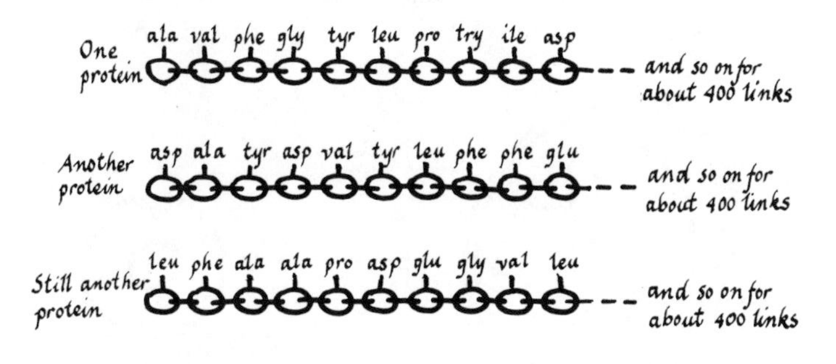

The amino acids are represented as chain links, with the first three letters of their names attached.

The twenty amino acids are: phenylalanine, leucine, isoleucine, methionine, valine, serine, proline, threonine, alanine, tyrosine, histidine, glutamine, asparagine, lysine, aspartic acid, glutamic acid, cysteine, tryptophane, arginine, and glycine.

Translation

Note how similar these five characteristics are to those of DNA (pages 15–16): chains, whose links are in a specific order; twenty different links (letters) in the protein alphabet and four in the DNA alphabet. It is immediately clear that the conversion of DNA information to protein substance must be essentially a language translation process: one sequence of letters in a four-letter alphabet to another sequence of letters in a twenty-letter alphabet. Again, somewhat like the translation of Morse code — a language with a two-letter alphabet (dot and dash) — into English, a language with a twenty-six-letter alphabet (a–z).

And that's just what happens. All cells contain thousands of tiny, incredibly ingenious translating machines that assemble protein chains. They are called *ribosomes.*

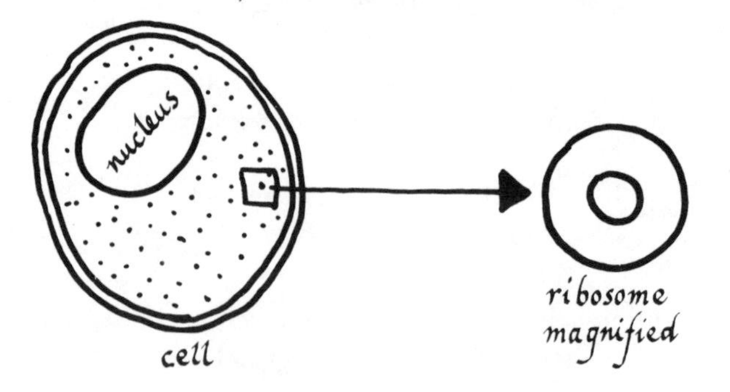

Here's how they work. First, a section of DNA's information — a gene — is copied by an enzyme — a protein that helps the process go faster. This gene copy is a chain called *messenger RNA* (messenger ribonucleic acid). RNA molecules are chain molecules almost identical to DNA, but not as long. One DNA molecule contains many genes; a messenger RNA molecule is a copy of just one gene. These particular RNA molecules are called *messenger* because they carry the gene's message from the nucleus, where the DNA is located, to the space in the cell outside the nucleus (cytoplasm), where proteins are made, on the ribosomes.

a segment of DNA about two genes in length
(2000 nucleotide links) represented as a line:

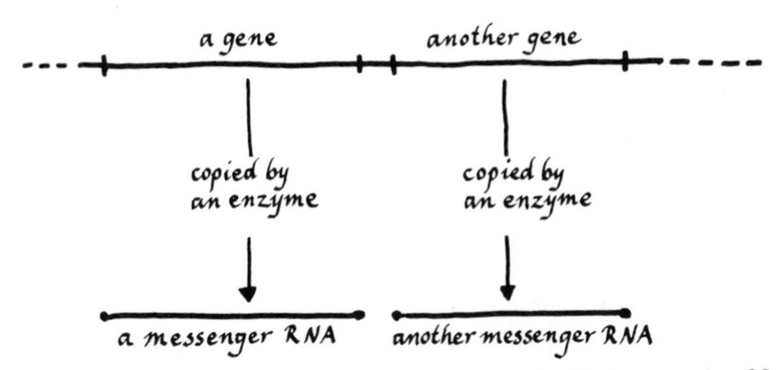

One end of the gene-copy messenger RNA inserts itself into the ribosome.

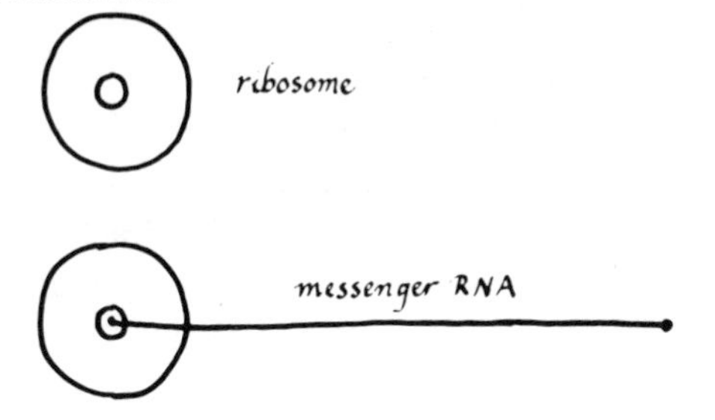

The ribosome is a reader; it reads the sequence of nucleotides (letters) in the messenger RNA, but instead of emitting speech it emits a protein. Here's how it's done. Special enzymes attach amino acids to a small RNA molecule called *transfer RNA* (tRNA). Each one of the twenty amino acids is attached

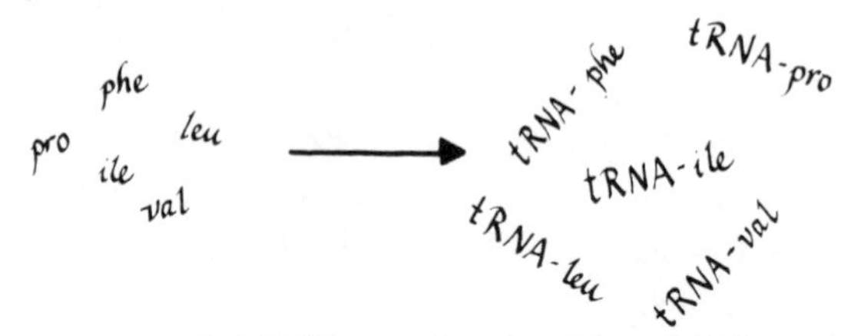

to its own special tRNA molecule. The tRNA molecules, with amino acid attached, present themselves to the ribosome.

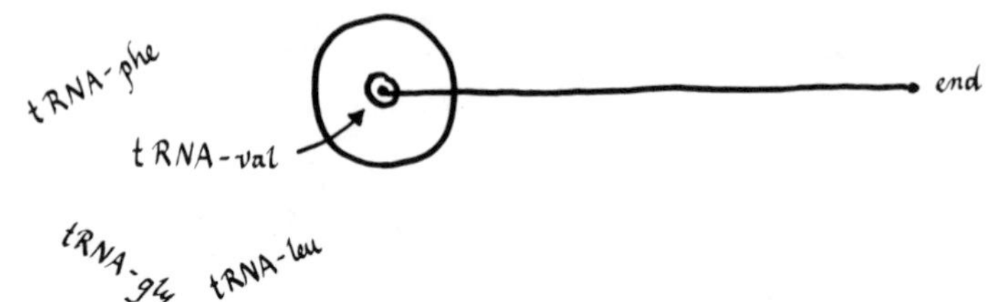

The ribosome selects the proper tRNA (with attached amino acid) according to the phraseology of the messenger it is reading at the moment. Thus, if the ribosome is reading on the messenger a group of nucleotides specifying amino acid ala (alanine), it selects a tRNA having a matching set of nucleotides to which amino acid ala is attached. The matching of messenger nucleotides to a particular amino acid, then, is dependent on the natural matching relationship of nucleotides. Each set of nucleotides on the messenger pairs up perfectly with a matching set of nucleotides on the transfer RNA. As each new amino acid and its tRNA comes into the ribosome and is properly positioned, the amino acid becomes chemically connected to the amino acid that preceded it into the ribosome.

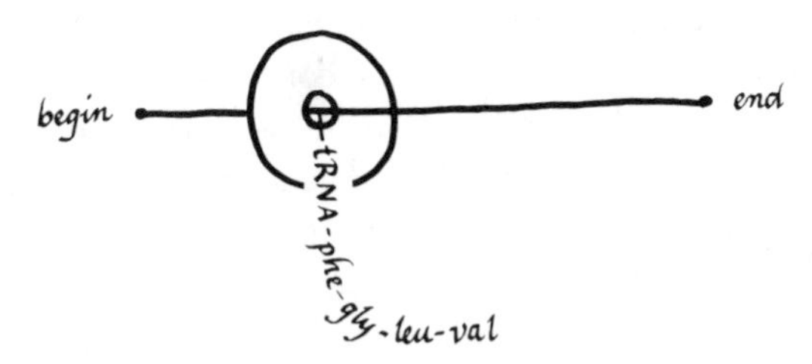

So the links are connected in sequence, one at a time. As the ribosome reads the messenger, the protein chain grows steadily in length.

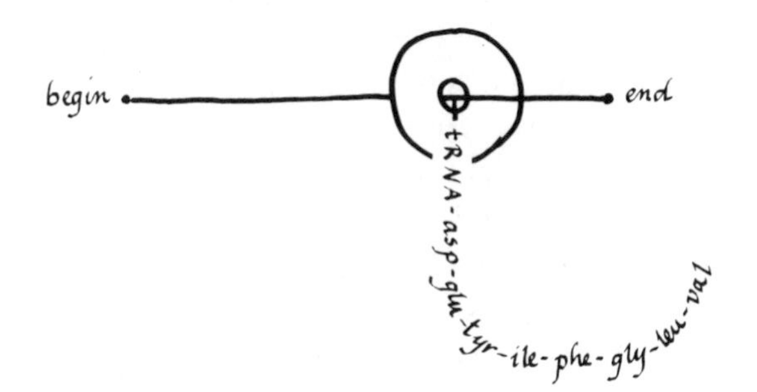

When the reading of the messenger chain is finished, the complete protein chain is released.

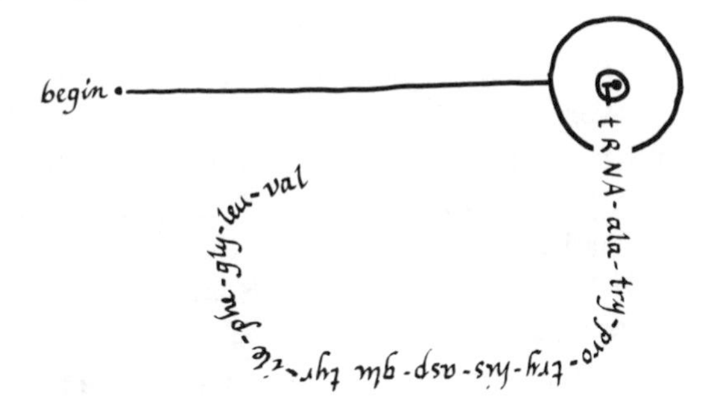

Thus is a new protein born. A sequence of nucleotides in one gene-length of DNA exactly specifies a sequence of amino acids in a protein. One gene; one protein. (The one gene–one protein concept was learned long before we knew how proteins were made. George Beadle, a biochemist working with bread mold in a series of brilliant experiments in the 1930s, had shown that changes in single genes resulted in damage to single proteins. This work was subsequently confirmed and extended, using bacteria. This great work — and much more that this story will tell of — would not have been possible without the important discovery, by Hermann Muller, in the 1920s that changes in DNA (mutations) could be produced at will by exposing living systems to x-rays.) The DNA contains as many genes as there are different proteins in the cell (2000 in the bacterium; 200,000 in the human).

The fidelity with which the protein-building machinery carries out this translation process is surely remarkable. Few mistakes can be tolerated in producing the thousands of proteins necessary for the cell's life. There is no machine humans have made that can transcribe the equivalent of 2000 novels with so few errors.

Discovery of tRNA

My former research mentor, Paul Zamecnik, and I jointly discovered transfer RNA and explained its mode of action in 1956. Zamecnik had already shown that ribosomes were the structures on which proteins were put together. I had proved the year before that amino acids are energized — that is, made ready to react with other amino acids — by a special set of enzymes. (This is described in Chapter IV.) But something was missing in between — something to which amino acids could be attached, thereby giving them the identity needed to "recognize" their place as prescribed by messenger RNA. Paul Zamecnik and I discovered in cells some small RNA molecules that had a marked affinity for amino acids; that is, amino acids became linked to them with exceptional firmness. We quickly recognized that we had fallen upon the missing intermediate in the construction of protein. By the end of the next year, after much intense and happy experimentation, we had a fairly complete picture of tRNA's mode of involvement in protein construction, as I have described it to you.

From Chains to Three-Dimensional Beings

So far the story is profoundly satisfying: living mechanisms use chains like language. To get from the plan to the finished product is a simple task of translation. But we still have one big hurdle to leap. Translation converts one symbol to another symbol; one dimension to one dimension; one chain to another chain; nucleotides to amino acids. How do we get from chain to *substance*; to protein molecules doing their jobs; to things we can touch and grasp — seeds, flowers, frogs, you, me? We have yet to make the leap from one dimension to three dimensions.

The answer lies in the nature of the links in the protein chains: the amino acids.

Though protein molecules are chains, they are in fact — physically — truly three-dimensional structures. Just as any real chain is. The twenty different amino acids of protein are not inert symbols; each has unique chemical properties. Some prefer making chemical bonds with their twins in a chain. Some are more acid; others, more alkaline. Some tend to seek water; others, to shun water.

Some are shaped in such a way that they tend to make the chain kink, and so forth. And several have very special properties to contribute to a protein to make it perform a unique function. Depending on where these amino acids are located in the chain, they determine the chain's ultimate shape. The chains, when completed, intertwine and fold upon themselves so as to become a sort of ball of string; more accurately, a ball of chain, a real, palpable, three-dimensional thing. The *sequence* of amino acids in the stretched-out chain determines the *exact* manner in which the molecule will fold when it is free to do so. The manner of folding, in turn, determines the shape, properties, and function of the protein molecule.

The gene for muscle proteins instructs the protein-building machinery to make a protein chain that will fold into a final shape, a long fiber having the ability to slide over its neighboring fibers and so cause contraction to occur. The protein chain that is to be the oxygen-carrying protein in blood cells, hemoglobin, folds into a three-dimensional shape, uniquely capable of holding and releasing oxygen. And so thousands of protein chains, each of whose sequences is determined by a sequence of nucleotides in genes, fold into special shapes and so gain unique functions.

Creating Order Is Mostly Making Chains

Remember what we said about order in the first chapter: life works toward order within a universe that is moving steadily toward disorder. Now we can see much more clearly exactly what this really means. Being alive means, literally, putting links in chains in an exact predefined order. Once the order is established, the acquisition of final form and function may be thought of as almost automatic; a spontaneous result, if you will, of putting one thing in front of another.

The Importance of Weak Chemical Linkages

A very interesting generalization has come from a study of the important molecules of the cell: DNA and RNA and protein. It is that *weak* chemical connections are vital to life. Strong linkages are the kind that connect amino acids together in protein and the kind that connect nucleotides in DNA and RNA. They are the bonds that hold each link of a chain to its neighbor. But weak linkages are the ones that cause and maintain the final shape and folding of all the large molecules. In DNA, it is weak linkages between nucleotides that hold two chains together to make a double helix, essential, as we shall see later, to DNA reproduction. In protein, it is weak linkages between amino acids that hold the proteins in the folded shapes essential to their function. And in the construction of new proteins on ribosomes, transfer RNA molecules are able to "find" their proper location on messenger RNA by matching up their nucleotides with complementary-shaped nucleotides on messenger RNA. The virtue of these essential linkages is that, by being weak, they are transitory. They serve their purpose and then can be easily broken and used again.

Viruses — Intimates of Life but Not Alive

Viruses are made of protein and either DNA or RNA. So they have information in the form of either DNA or RNA, and they have substantive identity in the form of protein.

But they cannot reproduce themselves without assistance. The help is provided by cells. The proteins of the virus help it find a cell and get into a cell. There it finds the machinery — the cell's machinery — to produce more of itself. It does so, and, having thus completed its task, it and its progeny work their way back out of the cell to repeat the whole unpleasant process in another cell. During this series of events the virus may kill its "host" cell, damage it, change it, or simply do nothing to it; it depends on the type of virus and the type of cell. An important change that a virus can induce in a cell is its conversion to cancer. This mysterious phenomenon is the basis of much present intensive effort in cancer research, as we'll see in Chapter VIII. We suspect that viruses, though simpler than cells, are not more primitive. It seems likely that they were once normal parts of cells that broke away some time in the remote past to establish their own parasitic form of "living." We do not think of viruses themselves as living, for they are not independently capable of reproducing themselves.

Mortality and Immortality

We now know that the creation of an individual requires a set of written instructions. The instructions have been copied over and over again with remarkable fidelity for many millions of years, yet the individual dies after only a few years. We may ask, then, if the instructions are immortal. Well, yes — at least as immortal as anything can be for the biologist. The fact is that the mortal, living-and-dying individual is the transient caretaker of the instructions that must be conveyed down the generations; the runner in the relay race where DNA is the baton. An individual life has significance only to the extent that it passes information about its ancestors to its descendants. Certain moths are born without mouths and begin to starve from the moment of their birth — their only job being to mate and lay eggs quickly so that moth information will get to the next generation.

If DNA is the mortal's immortality, the human's perverse curiosity can't resist asking: How did it all get started?

CHAPTER III

Beginning

Which came first, the chicken or the egg? This hackneyed query is unanswerable not because the answer plunges us into a monotonously repetitive cycle backward in time, but because, by virtue of the cumulative effect of tiny changes at each turn of the cycle, the protagonists eventually disappear! A chicken pedigree going backward a billion years would reveal a gradual transformation of our feathered companion into a form of life that would no longer be chicken, by any stretch of the imagination. I'd venture a guess that it would be smaller than the head of a pin and living in the ocean. We'd have much the same experience if we traced our own lineage far enough back.

And farther back still? We must assume there was a beginning. The last chapter's intimations of DNA's immortality need now to be put in better perspective. These enormous molecules, full of information needed to create our earth's present life forms, must have had a modest start some time in the distant past.

The best estimates are that life began some 3 billion

years ago, after the 2-billion-year-old earth had cooled sufficiently to support it. There are fossils of extremely small and quite simple sea creatures over 2 billion years old. Ancestors of those fossilized creatures would have been even smaller, and we can assume that the most primitive form of life of all was a cell, perhaps not unlike some of the simple forms of single-cell life that exist in abundance today.

So the focal question for us is: How could a cell have first got its start? Not how *did* a cell first get started — that's a question that can never be answered because no one was there to watch. But it's quite legitimate to ask how *could* it have happened. We can make astute guesses and do experiments that indicate probabilities.

The Essential Ingredients

We have a pretty good idea, based on conclusions of studies by geologists, paleontologists, physicists, and biologists, of what the world was probably like 3 billion years ago. Science-fiction books and movies picture it very vividly and probably accurately: an utterly barren, gray landscape of rock and lava, without a touch of green; erupting volcanoes; jagged mountain peaks; steaming seas; lowering clouds; and a steady, unending downpour of rain shattered periodically by streaks of lightning and blasts of thunder — seen and heard by no living creature. Certainly this would have been a miserable place for you and me. But it was a fine setting for the commencement of life. Here's what would have been needed to get things moving.

1. A warm temperature.
2. An ample supply of water.

3. Sources of the necessary atoms: carbon, hydrogen, nitrogen, oxygen, and phosphorus.

4. A source of energy.

No problem about water and warmth. As the earth was cooling, millions of years of rain had filled the oceans, which were heated by the still-warm earth. Lightning supplied energy aplenty, as did ultraviolet radiation from the sun whenever the sun broke through the clouds. (These rays were much more powerful then than now because the ozone layer above our atmosphere did not yet exist. Ozone is an effective absorber of ultraviolet radiation. The layer has been built up from oxygen that has been gradually accumulating on earth as the result of the activity of plant life.)

These conditions are certainly simple enough for simple beginnings: the making of information chains (DNA) and cell-substance chains (protein). But before we can have chains we must have links: the nucleotides of DNA and the amino acids of protein. As we know, these links are small molecules — arrangements of the elements carbon, hydrogen, nitrogen, oxygen, and phosphorus bound together chemically.

Creation of Simple Molecules

Here, then, is the scenario. Simple compounds containing carbon, hydrogen, nitrogen, oxygen, and phosphorus in the watery solution of the sea are constantly bombarded by lightning and ultraviolet light. They are thereby forced into various combinations, some of which are stable and durable.

As the process continues for hundreds upon hundreds of millions of years, the sea gradually becomes richer and richer in complex combinations of elements — molecules — including nucleotides and amino acids. Indeed, a time comes when the sea is a sort of soup, a pottage laden with a wondrous abundance of all kinds of new molecules.

The Importance of Time

Let's pause to grasp the significance of time in the process we're considering. The more time available, the more likely it is that anything that can happen will happen. So it is with chemical reactions: even the more improbable ones will occur if time is not limiting. And if the products of those reactions are stable, they would become relatively permanent constituents of sea water.

Soup Is Possible Because Life Is Absent

Now, if the idea of the sea being like soup seems far-out to you, it should. There is nothing comparable in our experience today. A rich brew like that can't possibly accumulate now because living things would eat it up. Bacteria and other small greedy creatures abound today, and whenever a good food supply appears, they consume it and increase their numbers until the supply is gone. So you can see that the oceans could become soupy because life was absent from it.

Laboratory Simulation of Ancient Events

Of course, what I've been describing is a hypothesis that can never be proved. We can, however, show in the laboratory that these things could have happened. It certainly ought to be possible to simulate in the laboratory the postulated primitive conditions. The simple compounds thought likely to have been present in the sea 3 billion years ago could be dissolved in water in a flask. The flask could be connected to a source of an electrical discharge to simulate the energy input from lightning. All parts of the system would have to be sterilized so that we could be certain no living cells were present. The electrical discharge could then be started and the contents cooked for a time. Then the flask could be opened and its content analyzed for newly created compounds.

I'm pleased to report that the experiment has been done and the result is completely convincing: both nucleotides and amino acids are created from the simple sources of the five elements! So the links of life's chains can be produced in a sealike environment by the use of lightning as an energy source.

Creation of Chain Molecules

The next step, obviously, is connecting links to form DNA-like and proteinlike chains. It is reasonable to suppose that if the laboratory simulation of primitive conditions promoted the formation of links, it could also impel the connection of links to form chains. And, sure enough, short chains do form. And they are identical in basic chemistry to today's DNA and proteins.

Again remember that these experiments show what could happen, not what did happen. The situation is not unlike the problem faced by Thor Heyerdahl in trying to establish whether the Polynesians found their present homeland by sailing west from South America. By successfully making the trip himself in a raft, he proved only that the journey could be made by someone using a similar mode of transport. He did not prove that the Polynesians had actually made the trip.

On to a Cell

From this point on we can envision five additional crucial steps upward to a cell.

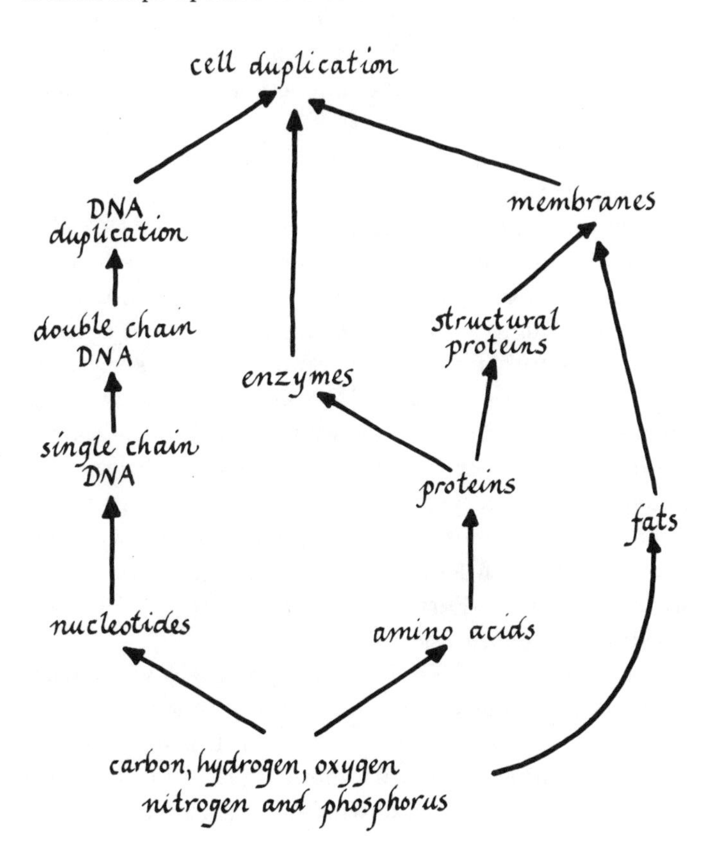

1. *Appearance of enzymes.* Enzymes are protein molecules — special kinds of protein molecules that make all chemical reactions in cells go faster. Every living cell today contains thousands of enzymes, each one doing its special job: breaking down food materials, producing energy from food, facilitating the construction of chain molecules from simple molecules, and innumerable other tasks. The slow pace of events associated with emerging life in the sea would be speeded up by enzymes. The earliest enzymes must have been short chains of amino acids that became linked together by chance. By repeated "trial and error," some of these combinations would have produced a protein with the unique ability to make some reactions speed up.

2. *The doubling of DNA.* You must try to envision millions of DNA chains throughout the oceans slowly growing in length by the random addition of nucleotides. Gradually, some meaningful sequences would be formed; meaningful in the sense that they would contain instructions for building a few primitive proteins. Some of these would be useful enzymes or parts of essential structures. The longer these delicate DNA molecules became, the greater was the danger they would break. So the chance occurrence of a means of protecting and stabilizing them would have been valuable. A simple doubling of the strands accomplishes this: two strands wound around one another are much less likely to be damaged than either strand separately. Furthermore, double-stranded DNA is essential to the process of DNA duplication.

3. *Duplication of DNA.* This is the process by which each strand of the doubled DNA chain makes an identical copy of itself; that is, a second double chain. It is an elegantly simple operation, begun, as shown here, when the two strands start separating, unwinding like a rope.

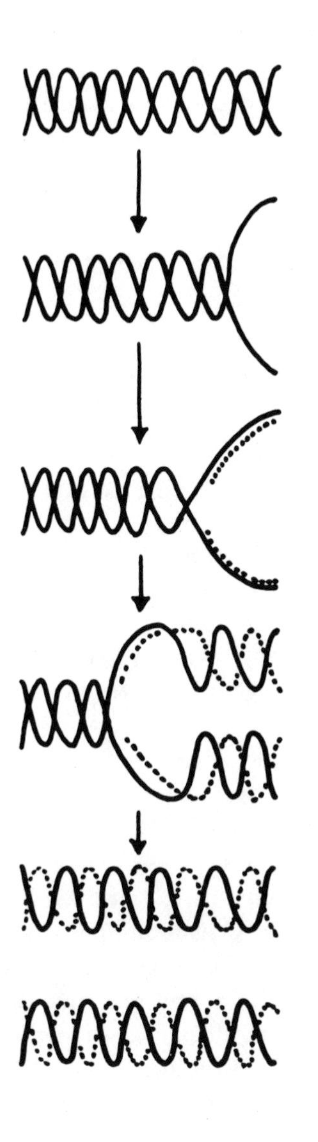
double DNA
unravelling
new nucleotides matching up with each strand
addition of new nucleotides continuing; new double DNA's forming
two new strands identical to old; each containing one new strand and one old

New nucleotides are lined up in order along each of the old strands and are then connected together. The sequence of nucleotides in each new strand is exactly specified by the sequence of nucleotides in the old strand. The reason for this is that the new nucleotides coming in to be connected will pair up only with their specific opposite number on the old strand.

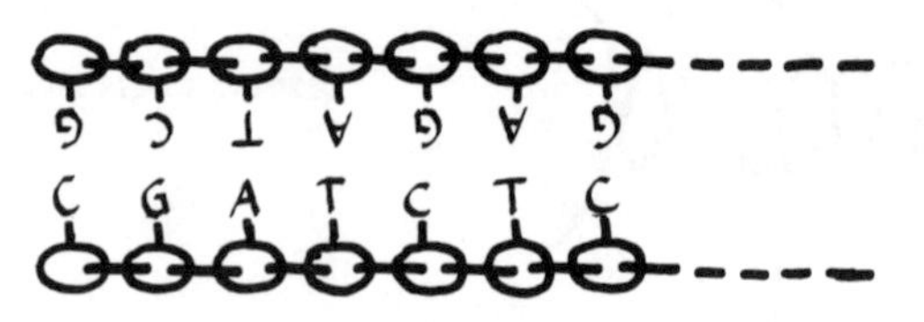

adenylic acid (A) always pairs with thymidylic acid (T)
guanylic acid (G) always pairs with cytidylic acid (C)

When the job is finished, there are *two double* chains, each an old chain paired with a new chain. One double chain is identical with the other double chain in every way. (When this task is completed in a cell, the cell is ready to divide in two. Each of the two new cells will get an identical double DNA chain.)

DNA duplication is not a job the DNA molecules and nucleotides can perform all by themselves. Now, and presumably at the beginning stages of life, the job, like all other cell reactions, requires enzymes.

4. *Envelopment of the key participants.* A critical event in cell genesis would have been the surrounding of the essential molecules by a wrapping, or envelope. This would have served both to protect the DNA and proteins and other critical molecules, and to bring them closer together so that they could work more efficiently in concert. Both proteins and fats, the latter providing the water-resisting quality that would ensure the cell's isolation from its

environment, are important elements in these cell-wrappings, which are called *membranes.*

5. *Cell duplication.* The acquisition by the molecules of an enveloping membrane would make something very like a cell. But a cell is a useless thing unless it can duplicate itself. The essential components are present: duplicatable information to make a new cell, and enzymes to carry out essential cell functions, all confined and protected within a membrane. Duplication of this whole package would obviously involve the coordination of all components in an extremely intricate operation of which we lack understanding even today. But once accomplished, the way to the present was open.

Life Arose Only Once

I speak as though this extraordinary genesis was a single event. Well, it probably was. There are two reasons for saying that. First, every living creature today — without exception — uses the same building materials (the same four nucleotides, the same twenty amino acids and other things) and the same general machinery for making its protein molecules (ribosomes, transfer RNA, messenger RNA) and for conducting the other business of life. We'd expect that if life had started more than once, each start would have involved different building materials and different machinery. The fact that all living creatures contain identical building materials and machinery strongly argues for a single origin.

Another reason for believing that life had only one origin is that the earliest form of life would have very rapidly consumed the soup from which it arose. Thus, it and its progeny would have eliminated the rich environment that millions of years of brewing had created. It may be a little hard to believe that a tiny cell and its progeny could rapidly rid the vast oceans of all that good food. But consider this. The common bacterium *E. coli* can double its number every twenty minutes if provided with its favorite food. This means that if we start right now with only 1 cell, we will have 2 in twenty minutes and 8 in an hour. We'll have 64 in two hours, 512 in three hours, 4096 in four hours, 32,768 in five hours, and so on. You can see that the increase in mass of cells is like an atomic chain reaction. In fact, if *E. coli* could continue to grow with plenty of food for *twenty-four hours,* the mass of cells would cover the earth a mile deep! The point is, a simple cell will eat and divide until its food supply is exhausted (or until its wastes poison it), and there's no limit to the mass of food that can be consumed. Thus, in a relatively short time, even if it were a slow divider, a primitive cell and its progeny could consume all the food in the oceans. Nothing would be left to permit any other form of life to emerge.

Introducing Energy

We have noted that energy was essential to the beginning of life. A flash of lightning, a burst of ultraviolet radiation, could cause molecules to connect up to make chain molecules, and this was, and is, the fundamental, essential building process of life.

In our little diagram on page 10, we showed that to create order among random objects we need information. In Chapter II we learned what the information actually is. We also saw that energy is required. We are now ready to explore in more detail the flow of energy in the living world.

CHAPTER IV

Energy

The radioactivity counter in my lab had just begun to print out the numbers I'd been impatiently waiting for. I had designed an experiment, as the result of a year's work, that would test an idea I had about how cells plugged energy into amino acids to make them link up to form protein molecules. To discover this would for the first time shed a clear light on the initial step in the building of the most important material of the body, protein. Well, those numbers, to my incredulity and dawning jubilation, fell out as though I had been in collusion with the atoms themselves, sharing with them my fervent hopes. Science teaches us to keep expectations low, but I had no doubt at that moment that I had made an important discovery. Not long after I published the work, it was confirmed by other researchers. It helped to open the door to a series of discoveries that within five years provided complete understanding of protein synthesis.

This chapter will be concerned further with these matters, but first we shall take a broader look at how plants and animals use energy, a subject we touched upon briefly in Chapter I.

A Willow Takes the Air

In 1630 Johann van Helmont planted a willow branch weighing 5 pounds in 200 pounds of soil. Five years later the willow branch had gained 165 pounds and the soil had lost two ounces! This experiment proved pretty convincingly that the soil was not the source of the bulk of material the plant was made of. Of course, water from the soil was essential to the plant; van Helmont watered his willow regularly and the plant used the water to help it grow. Even today, some find it difficult to envision where the material that plants are made of comes from, if not from the soil. The answer, that it comes from the air, seems hard to accept. But that's where the materials the plants need to construct themselves do come from — in the form of carbon dioxide. Water contributes hydrogen atoms to the building project and also accounts for a part of the plant's total weight. So we can see why van Helmont's willow could grow so exuberantly while taking so little from the soil.

Plants Capture Sunlight

But even with carbon dioxide van Helmont's willow tree would have languished without something else that soil or air or water could not provide: sunlight. The energy of sunlight was necessary to drive internal processes that would force carbon dioxide to become willow substance.

We noted in the last chapter that the energy that permitted life to get its start was probably derived from electrical discharges and ultraviolet rays. But in quite early stages of cell existence, a much more efficient device for securing energy emerged. This, the chlorophyll system, allowed plants to *trap* the energy of sunlight and put it to use inside the cell.

Chlorophyll accounts for the green color of the plants, grasses, leaves, and needles we're all familiar with. Chlorophyll is a green pigment, and its atoms are arranged in such a way that light, striking the surface of the plant, is captured within the molecule. With the help of nearby enzymes and other protein molecules, the chlorophyll molecules then convert the light first to electrical energy and then to chemical energy, and the latter can be used in plant construction.

A simple way of stating the total, worldwide accomplishment of plant life is:

$$\text{Light energy} + CO_2 + H_2O \rightarrow \text{sugar} + O_2$$

This formula says that plants, using the sun's energy, consume carbon dioxide molecules and water molecules and convert them to sugar molecules. Oxygen is discarded as waste. The sugar is used internally by the plants as a combustible energy source for the construction of plant substance. That is, plants eat their own sugar in order to grow.

Animals Consume Plants

We animals, of course, can't exist without free oxygen, and, as we have already learned, free oxygen was not available in the primitive atmosphere. But we see from the plant formula above that oxygen is a waste product of plants; they have no use for oxygen. As more and more plants appeared and took hold on earth over hundreds of millions of years, oxygen accumulated in the atmosphere, gradually creating an environment that could sustain animal life. Remember, too, as we noted earlier, that oxygen accumulating above the atmosphere gradually led to the buildup of the ozone layer, which now protects all of us, plants and animals, from the damaging effects of ultraviolet light. (The ozone layer is currently being damaged by fluorocarbons used as propellants in aerosol cans. Three billion of these cans are sold each year in our country. The fluorocarbons float up through the ozone layer, where they decompose and cause ozone molecules to be split to oxygen. The result is that increasing quantities of ultraviolet radiation are reaching the surface of the earth, where they damage the DNA of living things.)

At some stage of evolution, animal forms of life began to develop that "used" plants in two ways: they ate them for the food (sugar) they contained, and they absorbed (breathed) the oxygen they produced. The basic formula that describes the worldwide activity of animal life is, therefore:

$$\text{sugar} + O_2 \rightarrow CO_2 + H_2O + \text{energy}$$

Sugar, which animals obtain by eating plants, is "burned" in the presence of oxygen, yielding carbon dioxide and water as waste. This combustion process produces useful chemical energy inside the animal cells that can be used to construct animal substance. So animals eat plants (sugar) so that they may grow.

Plants and Animals Need Each Other

You'll see, now, that plants and animals are entirely dependent on one another. Plants make the oxygen that animals must breathe, and animals make the carbon dioxide that plants must use as building material. We can combine the plant and animal life formulas to make a cycle for all living creatures thus:

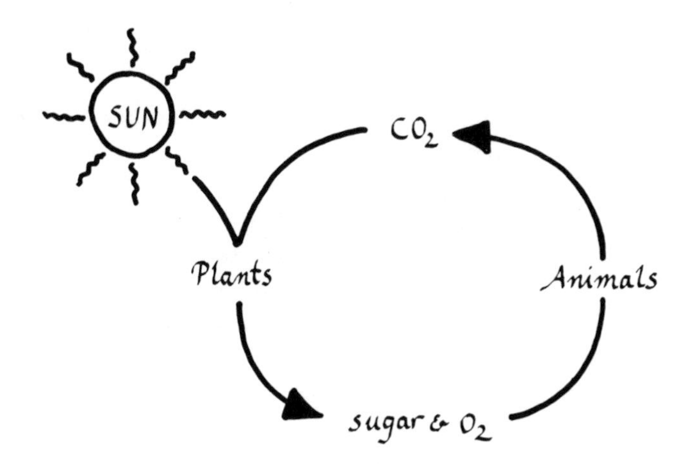

This cycle formula is simply another way of stating that the plant-life formula read backward is identical with the animal-life formula. It affirms the complete interdependence of plant and animal life, and the fact that plants would have had to become well established on earth before animals could get a foothold. It also is clear that *all* life — plant and animal — is dependent on the sun's energy; plants directly, animals through plants. Without a sun, our planet would be not only dark but very, very dead.

You can do a little experiment to show the interdependence of plant and animal. Seal a snail, a small water plant, and some water inside a test tube and expose it to the sun. Both snail and plant will be healthy for weeks. The snail eats part of the plant, produces CO_2 as waste. The plant consumes the CO_2, grows, produces O_2, which the snail uses to burn the plant's sugar it has eaten. Put the tube in the dark, and both snail and plant are soon dead.

Death Creates Life

When animals and plants die, the enormously complex organization of protein and DNA and RNA chains that is their bodies decays; that is, *other* organisms — mostly bacteria — feast upon the order created by life to obtain material they, in turn, can burn to make more of themselves. Carbon dioxide is the major waste product, and it returns to the atmosphere where new plant life can use it. Most of the carbon dioxide in the atmosphere comes from decaying plants and animals. If there were no decay, we would not have to worry about getting rid of carcasses, for all life would cease to exist in a few years.

Energy to Make Chains

It's easy to see what a resoundingly important evolutionary event the appearance of chlorophyll was. After it appeared, there was an almost explosive expansion of life — both plant and animal — on earth.

How is energy of the sun trapped by chlorophyll actually put to use in cells to accomplish the construction of cell substance? *The most important job to be done is the connecting of links to make chains*, as we now know. So we want to understand how energy makes the growth of chains possible.

ATP Is the Cell's Energy Currency

Light energy absorbed by chlorophyll is not, of itself, useful to plants. It must be converted to a useful form of cell energy, a form of chemical energy. The energy provider of all living cells, plant and animal, is *adenosine triphosphate*, better known as ATP. ATP is a small molecule, about the size and complexity of one nucleotide link in a DNA chain. It is, in fact, a nucleotide (adenosine monophosphate), with two additional phosphates attached to it.

Let's first examine how ATP is generated in the cell.

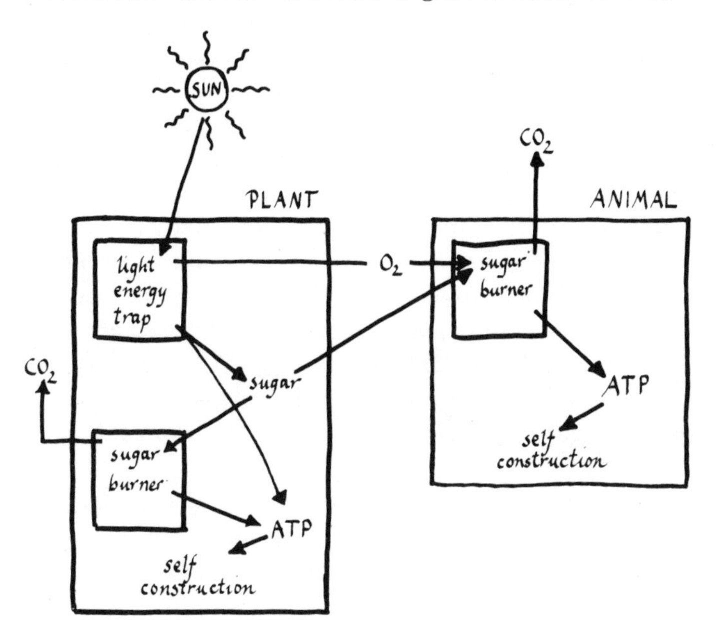

As you see here, the light energy trap (chlorophyll) absorbs light and converts it to electricity and so to ATP during the process of making sugar. The energy of the sun is thereby locked into, and so preserved, in ATP molecules.

Animal cells, of course, don't have any chlorophyll. They must generate ATP from the sugars they get by eating plants. Animal cells contain tiny sugar-burning chambers where plant sugar is burned in the presence of oxygen to generate ATP.

Combustion

The important difference between living combustion and ordinary burning is that in the latter the energy in the consumed material — whether it be coal, wood, or sugar — is released as heat. In the living process, the energy derived from the consumed material, sugar, is ATP. The making of ATP by burning sugar in animal cells has many similarities to the plant cell's making of ATP from light. Sugar-combustion generates an electric current; that is, a flow of electrons along a group of protein molecules. Similarly, after chlorophyll absorbs a portion of sunlight, electrons are generated and passed along a series of protein molecules like an electrical current. The current in both cases causes phosphate molecules to be attached to adenosine nucleotide, thereby creating ATP molecules. So, in both sunlight-capture and sugar-combustion, moving electrons are generated; these, in turn, make ATP molecules.

These two major jobs of ATP-manufacture are carried out in cells in special compartments that have their own enveloping membranes. They are almost like tiny cells within cells. *Chloroplasts* are where ATP is produced from sunlight in plants; *mitochondria* in animals are the sugar-burners where ATP is produced by combustion of sugar.

Plants Make Sugar for Themselves

From the foregoing you might conclude that plants make sugar to keep animals happy. This, of course, is not so. Sugar is the first main product of the photosynthesis process. But plants, like animals, need to burn that sugar to produce more ATP and other compounds for self-construction. They do this in sugar-burning bodies like mitochondria. So plants really have two energy-conversion mechanisms: one using solar energy to make sugar, and the other, animal-like, using sugar to make ATP and thence plant substance.

The Anatomy of ATP

Now we're ready to scrutinize ATP more carefully. To understand how it works, we recognize its key feature: it consists of a larger part, *adenosine monophosphate,* AMP, linked to a smaller part, pyrophosphate, PP. Thus, we write it AMP-PP, to show the two parts and the bond connecting them. AMP-PP, generated in the sugar-combustion process, *contains* energy stored in the bond between AMP and PP. It contains *potential* energy. We can show that this is true simply by breaking the AMP-PP bond to make AMP and PP.

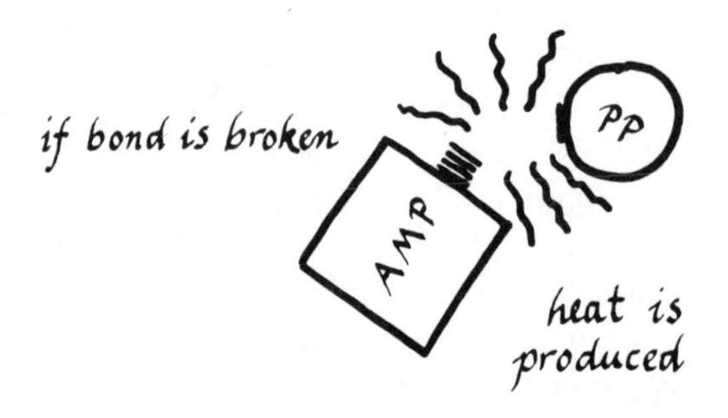

When we do this, there's a tiny explosion, releasing heat. But by far the most impressive way to show that AMP-PP harbors energy is to watch what the molecule does in living systems.

ATP Injects Energy into Links

I opened this chapter with a little of the flavor of my excitement at discovering how ATP's energy is used to build protein chains. Now we're ready to look at this essential first step in greater detail.

You can look at what's expected of ATP (AMP-PP) this way. If you have a handful of real wire chain links and you wish to connect them, you will get out a pair of pliers and proceed to open each link, hook it to the next link, and close the link again. You'll repeat this procedure until the job is done. You will have expended energy — physical muscular energy — to accomplish the task. ATP must do something analogous to what your hands and pliers do.

Here's the actual way ATP works to connect links in chains.

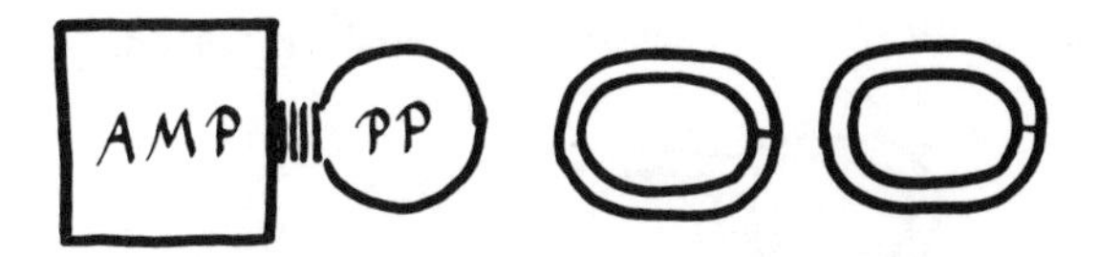

ATP and two links to be connected

I have pictured AMP-PP ready to interact with a real link, for example, an amino acid. In addition, I have shown two links of chain ready to be connected. I hope it is apparent that there is no way they can become connected without help.

The first essential step is an actual chemical attachment of the AMP portion of the AMP-PP to one of the links. When this happens, the PP is discarded. AMP has acquired the link at the expense of its partnership with PP. This reaction keeps intact the energy bond, which is now the connection between AMP and the link.

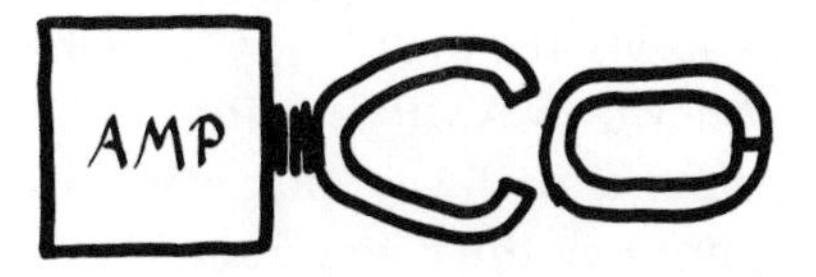

AMP attaches to link opening it

PP is discarded PP

The chain link is now said to be "activated," which means that it is altered so as to make it receptive to reaction with another link in the chain. I have pictured it as though the energy flowing from the new bond causes the link to open up. This activated state of the link is uncomfortable for it, and it "seeks" another link with which to react. That's shown in the final step, where a connection has been made, and the AMP is simultaneously released.

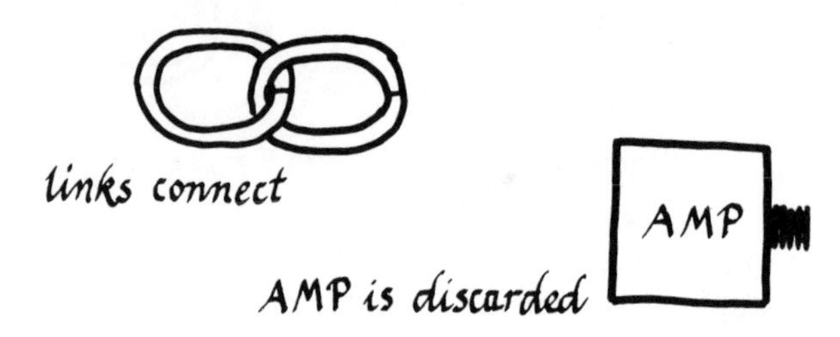

Note that the overall effect is that two separate links become two connected links, and ATP (AMP-PP) is broken down to AMP and PP.

Energy Is Carefully Conserved

Observe how carefully energy is conserved. If we took AMP-PP and broke it down chemically to AMP and PP, energy would emerge as wasted heat, as we've noted. The cell breaks AMP-PP down to AMP and PP — the same final result — but uses a pathway that conserves the energy in a building process: the connection of two links. Back in the mitochondria, AMP can have phosphates attached to it again to make more ATP. PP is split in two by a special enzyme and yields two phosphoric acids. These can be reused, too.

Nothing Without Enzymes

Now, these things can't happen without the help of enzymes. Enzymes make things occur at rates satisfactory for the cell's purposes. Enzymes, which are protein molecules, accomplish this activation by firmly holding the AMP-PP and the chain links in proper positions to ensure proximity.

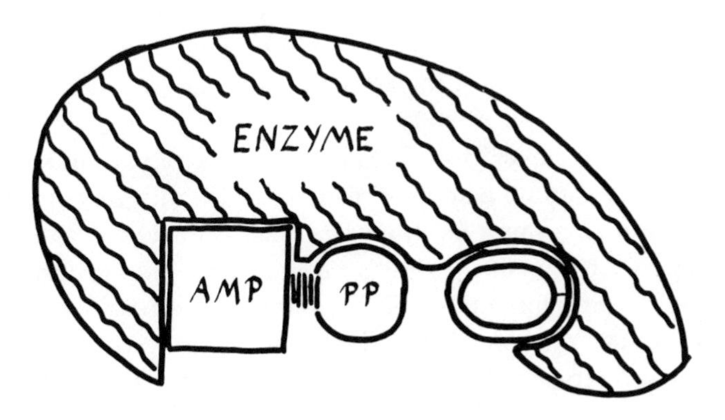

Once the participants are in the right relationship to one another, the rest of the events happen easily. Without enzymes, proximity of the participants would depend on chance and so would take an unconscionably long time.

Re-Enter transfer RNA

I've taken liberty with the true sequence of events to try to make clear to you how ATP participates in the protein chain construction process. Let me now set things straight. After an amino acid link is activated — "opened" — by becoming attached to AMP, it is not immediately connected to another amino acid link as I have depicted. You will recall from Chapter II that amino acids must be attached to tRNAs to give them an identity so that messenger RNA can recognize them for proper ordering. Only after the amino acids are placed in proper order on the ribosome or messenger-reading machine are they ready to connect with their neighboring amino acid. So in principle what I've described is right. It's just that the ready-to-react link — the "opened" link — is transferred to tRNA, still opened, and makes the connection with the next link only after it has been properly positioned on the ribosome.

Note how clever the mechanism is. The activation step makes the amino acid link ready to react with *any* other link. But this won't do; the links must connect only after they're arranged in the proper order. The ordering requires attachment of each amino acid to its specific tRNA. The enzyme that activates — opens — the amino acid also attaches it in its opened state to the right tRNA.

So we now know the way energy is used to put links together in protein molecules in all living systems. As we've said, protein is the main material substance of life and it therefore represents an enormous mine not only of information but also of energy because each linkage preserves the energy derived from ATP. Similar principles are used to connect the links of DNA and RNA molecules, and of other types of molecules.

ATP Makes Everything Go

Finally, you must gain some sense of how all-pervasive ATP is in living systems. It is truly the universal unit of energy exchange. We have limited our discussion to its use in building chain molecules. However, in an animal like ourselves, only about 10 percent of our daily use of ATP is for this purpose. Most of the rest is used for making muscles move. ATP supplies the energy so that muscle filaments can slide past one another, thereby causing contraction. Other processes that involve movement — such as the conveyance (transport) chemicals across cell membranes — require ATP. But to me one of the most appealing uses of ATP, and one that illuminates well its mechanism of action, is its igniting of the firefly's lantern. The emission of light — any light — requires energy. In a flashlight a battery supplies it; in a firefly ATP supplies it. If you grind up some firefly lanterns in water, discard the particulate material and put the clear solution of lantern proteins in a test tube, hold the test tube in a dark closet, and add a tiny amount of ATP, the whole tube lights up! After the glow subsides, more ATP will ignite it again, and so on, indefinitely.

The study of this system reveals that the mechanism is essentially identical to the one I have described for connecting chains. ATP attaches its AMP part to a protein, and the energy transferred to the protein causes it to change shape. When this particular protein changes shape, it emits light. My colleague William McElroy discovered these things with the help of hundreds of boys, who were paid a penny for each hundred fireflies they captured. That was twenty years ago. Today we can purchase bottles of firefly tails (lanterns) from laboratory suppliers at small cost.

Life on Mars?

This may seem an odd point in our story to leap to another planet. But, as I write, we Earthlings have just landed a Viking vehicle on Mars and its arm is reaching out to obtain soil samples to test for life. Furthermore, you, the reader, now have enough background to understand how we might go about looking for clues of life on Mars.

Conditions on Mars are such that whatever life is there will not be visible to the naked eye; it will be microscopic and will have to be detected in a relatively small sample of surface soil that the scoop of the lander can grasp.

You should now be able to make a pretty good guess as to what we should look for, particularly if I tell you that there's plenty of carbon dioxide in Martian air. We should look for something like a primitive plant that could, in the presence of sunlight, convert carbon dioxide to some more complex material. The tiny laboratory in the lander can detect this. It gives radioactive carbon dioxide to a sample of soil and then tests the soil to see if the radioactivity has become a part of any larger molecules. Or, we can look for animal-like life by adding some radioactive sugar and seeing if radioactive carbon dioxide comes out of the soil. This would mean that something in the soil may be able to burn sugar.

The present Martian laboratory is capable of doing these and related automated experiments and transmitting the results of its work back to us on earth. It seems most unlikely that these first experiments will be conclusive, but they point up the value of our knowledge of life on this earth in determining how to look for clues to life elsewhere.

We have learned how information and energy work together to accomplish the fundamental job of building chains. We've learned how these processes, as we now understand them, might have originated on earth. We must now explore the underlying forces nature has applied since the simple beginning to produce the elaborate complexity of life we find on the earth now, 3 billion years later.

CHAPTER V

Change

During the whole brilliant period of discovery of the principles of life I have described so far, genetic and evolutionary thought and practice in the Soviet Union was controlled by a charlatan. T. D. Lysenko, an ineffectual scientist but a vigorous polemicist, persuaded first Stalin and then Khrushchev that acquired characteristics could be inherited and that scientists who believed otherwise should be silenced. And so they were, from the mid-1930s to the mid-1960s. During this period the government, following Lysenko's theories, tried to adapt tropical plants to grow in the Arctic and to force winter wheat to grow in spring wheat regions, raising havoc with Russia's agricultural production. Lysenko's concepts, incidentally, completely rejected the role of DNA as the material basis of heredity.

I was reminded of this ludicrous monument to wishful thinking as I watched a basketball game one day. Unbidden, Lysenko suddenly popped into my mind. Would he have ordered us to believe that basketball players are *stretched* by the game; that they produced taller offspring as a consequence? If each new generation played basketball, would the hoop have to be moved steadily toward the ceiling?

The popularity of this concept of the inheritance of acquired characteristics with the Soviet government is understandable, for it meant crop yields could be improved if the seed were conditioned. And it appealed to the human desire to influence one's own development, and pass accomplishments on to one's children.

But reality is harsh. The heights of professional basketball players are the result of a completely random, chance process that makes some people taller than others. Then the tall are selected as best suited to play basketball.

In this chapter and the next we shall focus on change and choice: on the origin of difference and variation in living beings, and the way in which the environment chooses the forms of life it will permit to survive. Throughout the long history of life, flexibility and adaptability — changeability — have been treasured assets. In the generally hostile and forever-changing environment of the earth's surface, change and survival were synonymous.

We are, of course, speaking about *evolution*, the process that has brought you and me, chicken, egg, basketball player, and all the rest of earth's creatures from that humble single-cell beginning to the present.

There are two means by which living things are changed in evolution: *mutation of DNA* and *sexual mixing of DNAs.*

Changing the DNA of an organism, as we now know, will result in change in the organism itself. The fate of the changed organism, then, depends on how it manages in its environment. This, the process of selection, will be examined in the next chapter.

Mutations

A mutation is a change in one or more of the four nucleotide links in DNA.

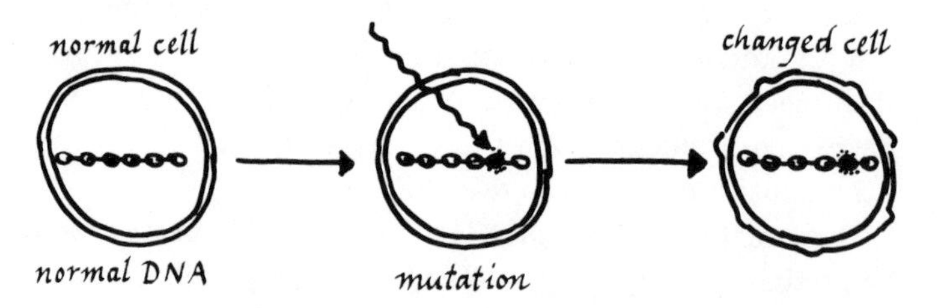

A change in even one link means, as you'll recall, a change in one letter of DNA's message. The messenger RNA copied from that DNA will contain the change and will be read differently by the machinery that is making protein. The result will be an altered protein; one amino acid link in the chain will be different. As a consequence, the function of the protein will be changed.

A very important feature of mutations is that they are copied when DNA is copied. Preparatory to cell division, as we have already explained, an enzyme copies DNA, nucleotide by nucleotide, until a completely new set of genes is reproduced. A mutation in the DNA will frequently be copied so as to perpetuate the error and spread it to all subsequent generations of cells containing that DNA. Thus is one tiny mutation forever inscribed in DNA's language.

Cause of Mutations

Mutations are caused by natural radiations (such as x-rays and ultraviolet light) and by manmade chemicals hitting and damaging the nucleotide links of DNA. Nucleotides may thus be transformed to other nucleotides,

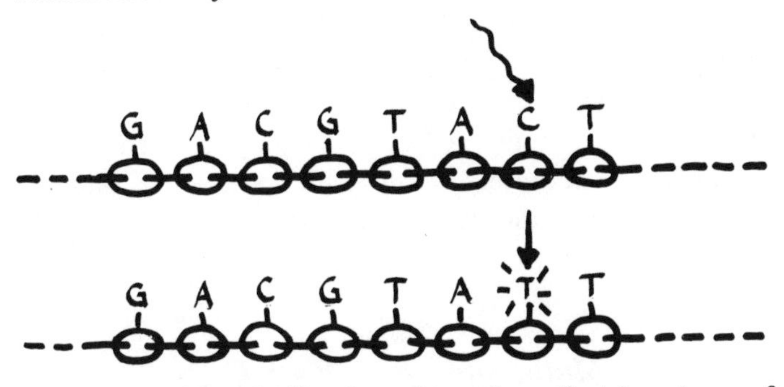

they may be chemically altered to a form that is not one of the four standard nucleotides,

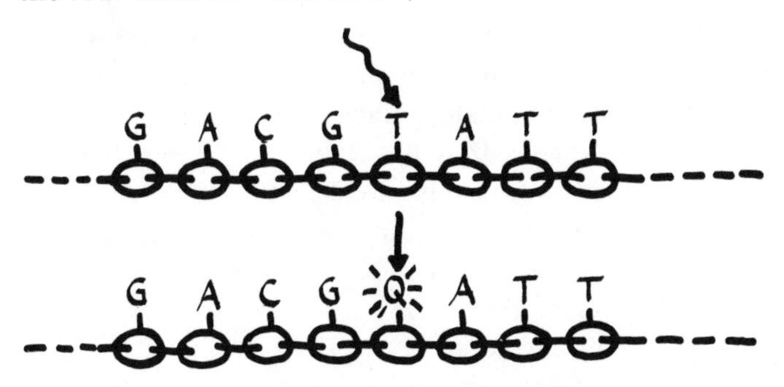

or they may even be taken right out of the chain.

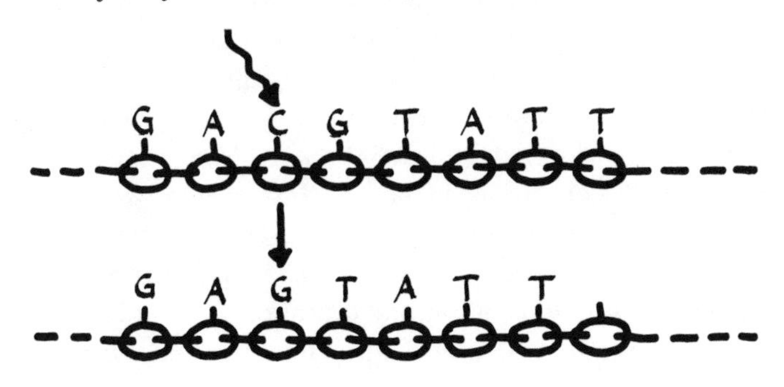

All these alterations naturally change the meaning of the chain; the language is thenceforth slightly different.

Mutations are purely *chance* events. There is absolutely no way of knowing which links of DNA will be hit. Mutations can happen at any time to any nucleotide of DNA of any living creature, including us. (There are remarkably accurate enzymes that constantly monitor DNA, and when they find a change, *repair* it! But they can't catch everything.)

Mutations Affect Body Cells and Sex Cells Differently

All of our body cells contain two entire complements of DNA: one from our mother and one from our father. In order for parents to make children, they must place their DNA in single cells that have the unique capability of mating; that is, uniting with a cell of the opposite sex and so sharing their DNAs. These specialized single cells are sperms, made in the testes of males, and eggs, made in the ovaries of females.

When mutations occur in the DNA of one of our body cells — one that is not growing and is one of billions of other cells making up our total structure — we most likely never know of it. Damage to one cell among billions can hardly be felt. (There is one important exception: the mutation that causes the cell to become cancerous. We'll consider this change in a later chapter.) However, when a mutation occurs in cells of testis or ovary, which is producing the sperms or eggs to be used for making new individuals, the situation is quite different. For if a sperm or egg contains a mutation, the mutation will naturally be carried into the fertilized egg. When the fertilized egg divides, the mutation will be copied into all descendent cells. Thus, *every single body cell* of the resulting adult will have a copy of the mutation. And every sex cell of this adult that will make its sperms or eggs in testis or ovary will also carry the mutation.

So *the kind of mutation that's important in evolution is the one that occurs in an organism's sex cells and is thereby inherited.*

"Good" Mutations and "Bad" Mutations

Mutations are rare but have, nevertheless, been essential instruments of change in evolution. They've produced changes in an organism's proteins, which gave the organism some advantage in coping with its surroundings. In this sense mutations have been beneficial. And there's no reason to believe that this process isn't still going on; that is, DNA's being altered so as to give now-living creatures, including humans, some slightly improved proteins that will make us function better. Of course, beneficial mutations are not easy to discover. A mutational change in DNA that alters an amino acid in protein so as to make the protein work better is likely to go undetected. There's simply no easy way for us to measure small improvements.

But by far the majority of mutations, at least the ones we can detect, are harmful. In contrast to beneficial mutations, mutations that produce bad effects are easy to detect because they are manifested as a defect, a weakness, a disease. We are learning almost daily of new diseases that are caused in humans by mutations. There are a very large number of these diseases, each one quite rare. In each case the basis of the disease is a mutation in sex-cell DNA that has been passed, in sperm and egg, to the next generation. In that generation all cells of the body have a copy of the mutation. One example, the most thoroughly studied, is sickle cell anemia. Here the mutation in DNA occurs in the gene that specifies protein molecules called *hemoglobin.* These are the proteins inside our red blood cells that carry oxygen from our lungs to the cells of our body. The mutation in DNA, copied into an altered messenger RNA, causes defective hemoglobin molecules to be made by the red blood cells — molecules in which a single amino acid in every chain is changed. This one change causes the hemoglobin molecules to change their shape. These many misshapen hemoglobin molecules inside the red blood cells strain the membrane, making the cells themselves misshapen, sometimes resembling sickles. The deformed red blood cells break, stick in blood vessels, and may cause serious illness.

Perhaps it should not be surprising that most mutations are detrimental. Look at it this way. The information that resides in organisms that are alive today — the cumulative work of 3 billion years of evolution — is far more refined than the work of all the world's great poets combined. The chance that a *random* change of a letter or word or phrase would *improve* the reading is remote; on the other hand, it is very likely that a random hit would be harmful. It is for this reason that many biologists view with dismay the proliferation of nuclear weapons, nuclear power plants, and industrially generated mutagenic (mutation-producing) chemicals. The earth's store of DNA is an immeasurably precious commodity — irreplaceable because evolution can never be repeated. Damage to the work of 3 billion years of evolution would be a monstrous atrocity, far worse than the destruction of all the works of the world's great artists.

Sex-Mixing of DNA

So mutations are chance, indiscriminate events, mostly harmful — though providing sufficient means of changing DNA to have value in evolution. But something better was needed; something that would change DNA without the risk of damaging it. Ideally, *changing DNA by mixing it with a different DNA would accomplish this purpose.* If two different cells could get together and agree to recombine their DNAs so that gene-lengths

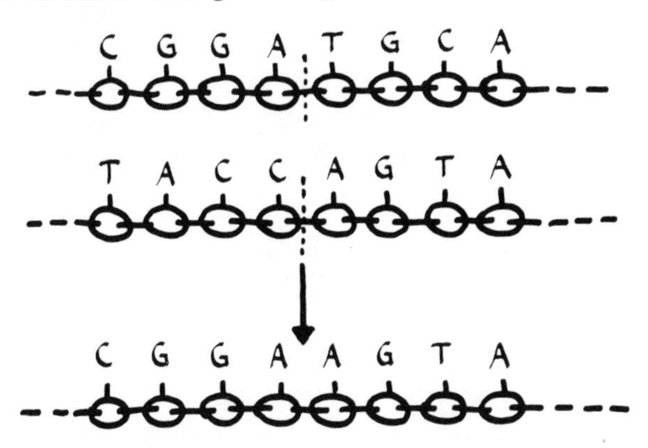

of one DNA could become directly linked to gene-lengths of the other, thereby creating offspring with genes from both, a giant step forward would have been made.

The great advantage of this method of changing DNA was that the change was accomplished by introducing a segment of DNA (gene) that had already proven itself, in evolution, to be successful. The change would then generally be a "good" one because both DNA participants would have been proven by the test of evolutionary success, survival. In contrast, a mutational change in DNA is more frequently damaging than beneficial, as we have learned.

Cell Fusion

At some time in the very remote past some kind of random bringing together of two cells' DNA must have happened. Presumably, cells back in that ancient soup experienced, in their frequent contacts, occasional unions that stuck long enough to allow merger, or *fusion*, of cells to occur. This means that at the point of contact the membrane envelopes of two cells broke down so that the contents could mix. The behavior of the resulting single larger cell would then be governed by the proteins of *both* cells. And both cells' DNA would be within one cell.

Something much like this actually happens today, as is shown in the next figure. Cells from different parts of the body, or even from different animals – for example, humans and chickens – can be brought together and induced to fuse in the laboratory. The resulting cell contains a double amount of DNA, and when it divides, the progeny often contain the double amount of DNA.

The Birth of Sex

The fact that cell union *can* occur suggests that it could have been an early mode of getting one cell's DNA together with another cell's DNA. Whatever the first method of mixing of DNA, it can be viewed as the beginning of sex. And it was surely one of the most momentous steps in evolution – of a significance comparable to the emergence of chlorophyll, with its ability to capture the sun's energy. For it explosively amplified life's potential for change, adaptation, and diversification.

The next phase, probably occurring while cells were still in the soup and still single, was to make the procedure more likely to occur; that is, to make contact between cells

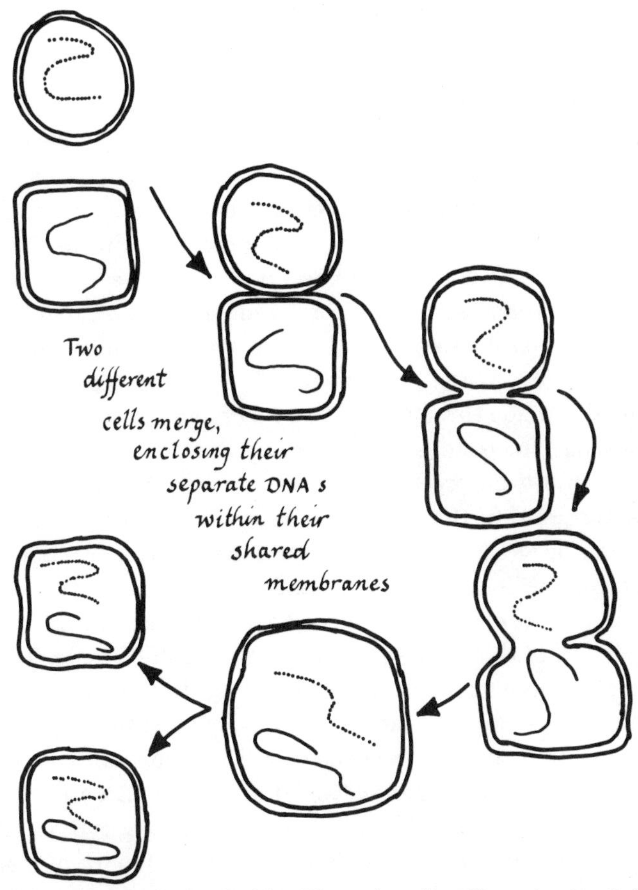

more specific and reliable. Though cell-cell contact might still be a chance event, if some cells fit snugly up against other cells by virtue of a complementary shape of their "skin," they'd be more likely to stick together. The appearance of complementarity in cell-cell interaction meant that two distinct populations of cells arose — one having an elementary character of maleness; the other, of femaleness.

Bacterial Mating

The counterpart of this phenomenon is sexual mating of bacteria, as we see it today in the laboratory. This is probably very much as it was in the early stages of evolution. Certain bacterial cells have the quality of male, or donor; others, the quality of female or recipient. The membrane surfaces of the two types are complementary and so mutually attractive (for example, if one has bumps, the other has depressions as can be seen in the illustration on page 69). When cells of opposite sex get together, a tunnel is made through the touching membranes, thereby bringing the insides into contact. The male, or donor, bacterium then proceeds to inject its pure DNA through the tunnel into the female, or recipient bacterium. This process occurs at a steady rate and takes about two hours to complete — to get all the male DNA into the female. Generally, the mating breaks off spontaneously after only a portion of the male's DNA has entered the female. The female's DNA remains intact inside her. In the laboratory we can separate the mating partners by vigorously mixing the solution they're suspended in. Thus, one can control exactly how many genes from the male will enter the female! The important point to make, however, is that the genes that do enter the female become a part of her information store. The male genes are incorporated right in line with the female's genes. When she later divides by cell-splitting, the daughter cells contain the genes from both male and female — as do all future generations.

Thus was the dawn of sex. Bacterial mating certainly illustrates vividly the purpose of sex: to recombine DNAs from different sources.

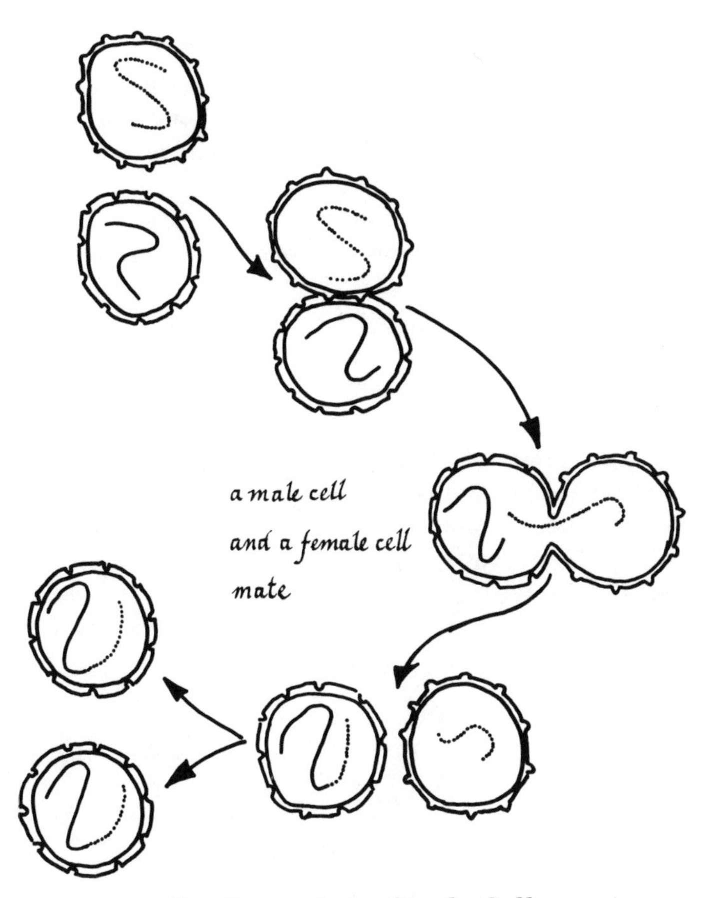

Sex Beyond the Single Cell

As organisms grew in complexity beyond the single-cell stage, the mixing of DNA by simple fusion of two cells became impossible. It was necessary to devise special means of sexually recombining DNA.

The objective remained the same: get a single cell containing the organism's DNA together with a single cell containing the other organism's DNA. Special organs evolved, the ovary in the female and the testis in the male, and each of these organs produced single cells. Like the bacteria, the cells had to be structured so that when they met they'd fuse. Their DNAs could then merge into a single cell — the sperm-fertilized egg.

Now, you'll recall that all body cells contain identical double sets of DNA — one set from the mother and one from the father. If sperm and egg each carried two complete sets of DNA (one from each parent) as body cells do, the union of sperm and egg would produce a new individual with four complete sets of DNA in its cells — an impossible situation. Obviously, then, it is essential that the amount of DNA that body cells contain be reduced to half before the DNA is packaged into sperm and egg. This vital process is accomplished in the testes and ovaries: here cells, containing the standard two sets of DNA (one from the father and one from the mother — mix, or shuffle, the two parental DNAs to make two DNAs in each of which genes of mother and father are evenly mixed.

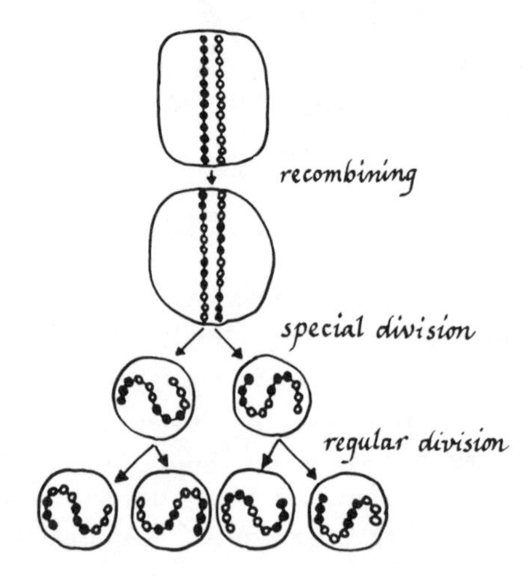

Then this cell divides, one mixed DNA going to each of the offspring. These offspring are the sperm in the male and the egg in the female. Thus, each sperm and each egg gets one complete DNA set in which mother and father traits are randomly mixed. Then, whenever sperm and egg unite, the resulting fertilized egg cell again contains two full charges, or sets, of DNA.

Since this vital process is not easy to understand, let's go over it again, with the aid of a diagram, beginning with union of sperm and egg — fertilization.

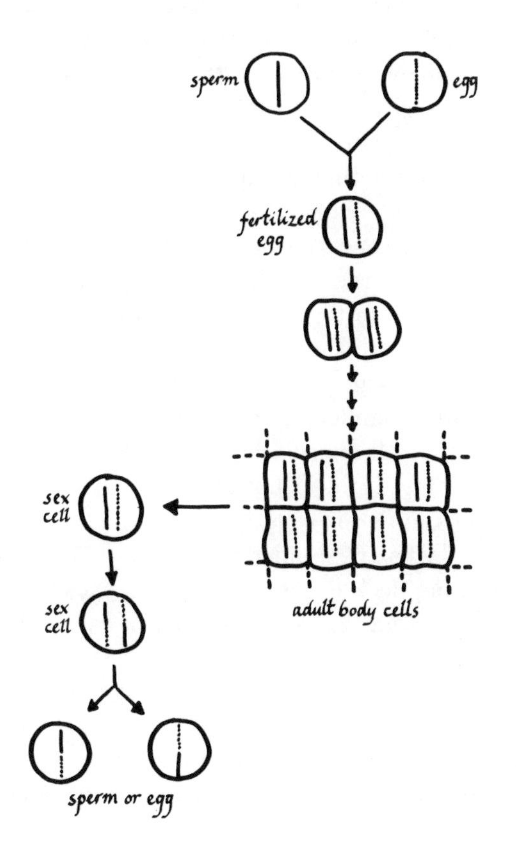

1. Sperm and egg each contains *one* complete set of genes.

2. Sperm and egg unite to make a fertilized egg, with one complete set of genes from mother and one complete set from father.

3. The fertilized egg divides, and cell division continues until a complete adult is made, many billions of cells, all with the same double set of genes that the fertilized egg had: one set from mother and one from father.

4. In the sex cells of the testes and ovaries of the adult (a) the mother's set of genes is mixed with the father's set of genes, and (b) the sex cells divide in a special way, leaving one mixed — mother, father — set of genes in each cell. These cells develop into sperms or eggs, ready to start the cycle again.

It occurs to me that another way of envisioning the above is:

1. Sperm and egg each contains a complete deck of well-shuffled playing cards. The decks are identical, though the order of cards from top to bottom is different in the two and is equivalent to the order of bases in DNA.

2. Sperm and egg unite to make a fertilized egg cell, which has two complete and separate decks.

3. Repeated cell division leads to billions of body cells, each containing two decks of cards.

4. In the adult's sex cells:

a. the decks are shuffled to make *one* 104-card deck;

b. the big deck is cut to make two packs again, and the sex cells divide, putting one pack in each cell.

Other Ways to Mix DNA

It seems appropriate, following a section on evolution's methods for mixing DNA, to say something about techniques that we scientists have discovered for combining DNA. Recombinant DNA research, the art of recombining DNAs from different organisms in living cells in order to obtain large quantities of the genes for study, is receiving increasing public attention.

Here's what's involved. Bacteria contain, in addition to their major piece of DNA (chromosome), another smaller molecule of DNA called a *plasmid.* Plasmid DNA is circular in conformation rather than straight, and it can get in and out of bacteria quite readily. These two properties of plasmids are put to use in recombinant DNA research.

If plasmid DNA is exposed to a certain enzyme, the circle opens up. (In the figure on page 74, the plasmid is much exaggerated in size.) Then, if another piece of ordinary straight DNA, from any source, is mixed with the opened-up plasmid DNA, the two DNAs join together. In fact, a larger circle is formed, containing the plasmid DNA and the other DNA. If the other piece of DNA, for example, came from human cells, we would have a bacterial plasmid containing a piece of human DNA. This is what is meant by recombinant DNA. The hybrid plasmids so produced can be made to re-enter bacterial cells, and as the bacterial cells continue to divide and increase in number, the plasmid, with its "foreign" genes, will be multiplied, too.

The objective of the scientist who uses the technique is essentially similar to any chemical "scaling-up" procedure: the obtaining of large amounts of particular genes to work with. Many biologists look upon the method as one of the most valuable tools to have been discovered in biological research. It will undoubtedly play a key role in unraveling gene control in embryogenesis and aberrant gene control in cancer. It also has potential for the large-

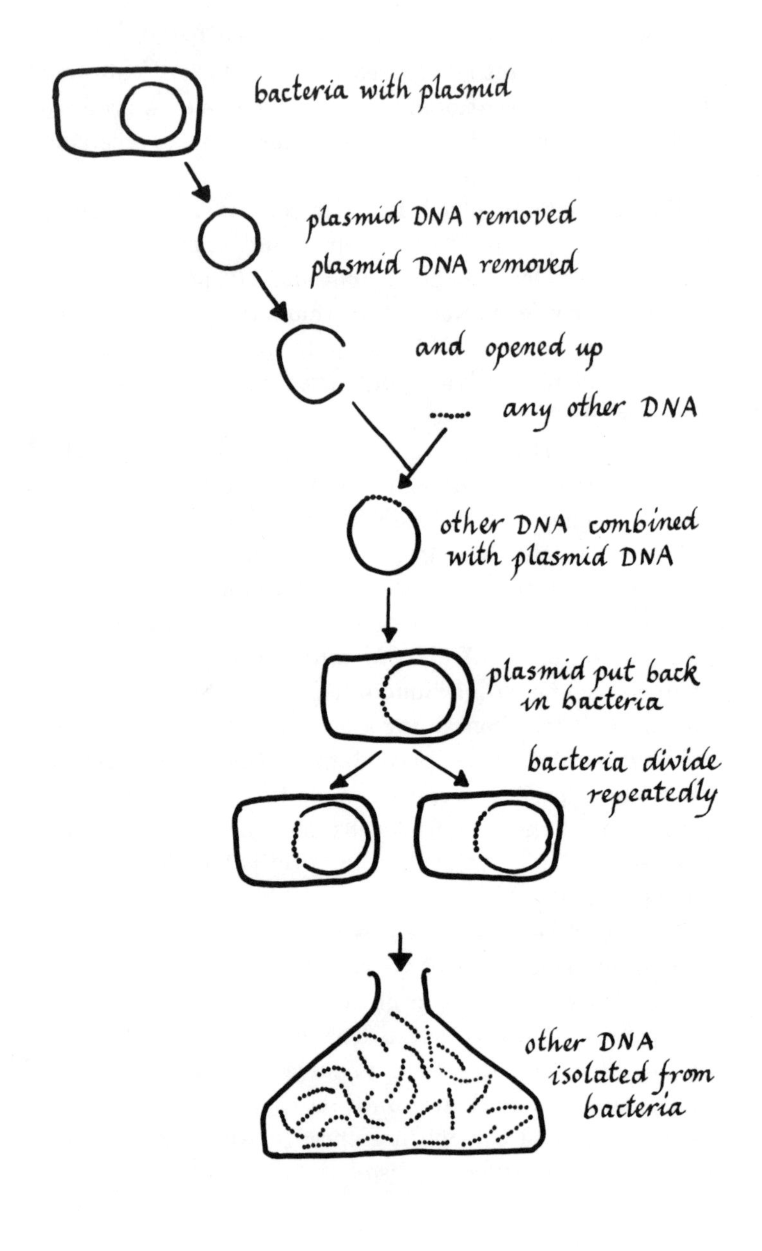
bacteria with plasmid
plasmid DNA removed
plasmid DNA removed
and opened up
any other DNA
other DNA combined with plasmid DNA
plasmid put back in bacteria
bacteria divide repeatedly
other DNA isolated from bacteria

scale production of protein gene products needed in medicine, like insulin. More remote is the possibility of using bacterially reproduced human genes for gene-replacement therapy in patients with hereditary deficiencies of particular genes.

Evolution is the history of ever-expanding *diversity*. Constant mutational changes in DNA, constant sexual recombining of DNA, created many differences among individuals. As individual differences continued to accumulate, sexual mixing could occur only between similar organisms. Thus, new species were launched to evolve separately. Thereafter sexual mixing of DNA occurred only between members within a species. In this way each species preserved a set of genes useful to it, and avoided becoming overburdened with a growing collection of extraneous genes. But it should be remembered that differences among individuals of the same species, and differences among species have the same underlying basis – changing DNA.

Now, we must recognize a very large qualification about change. Change in a vacuum is meaningless. An altered organism's success or failure can be measured only in relation to how the environment "views" the change. The environment is the judge of the significance of biological change. In the next chapter we will put together mutation, sex, and environmental selection to complete the picture of the evolutionary process.

CHAPTER VI

Selection

If a fat person and a thin person, of the same age and general health, both fell off a boat in the North Atlantic, the fat one would be more likely to see dry land again. There are two reasons for this. Fat is excellent insulation against the cold, as the bodies of whales, seals, and the like attest. And, since fat is lighter than water, it helps a swimmer to stay on the surface.

The lesson here is that the value, or usefulness, of particular traits of an organism can be assessed only in relation to the environment in which it finds itself. Although carrying around a load of fat would, in most situations, be disadvantageous, when one is dropped with it into the North Atlantic the sea becomes the judge of the value of obesity. The Atlantic's verdict is that fatness has survival value; it's good to be fat.

Environment and Change

Our fat and thin sailors help to dramatize my point, but really to get down to the nub of the relationship between difference and environment we must think in terms of populations and their progeny over many generations. If parents living in a particular environment pass a changed DNA to their children, those children, *their* children, and all succeeding generations will get along (1) as well as the parents, (2) better than the parents, (3) worse than the parents. This is shown more schematically with some dividing cells.

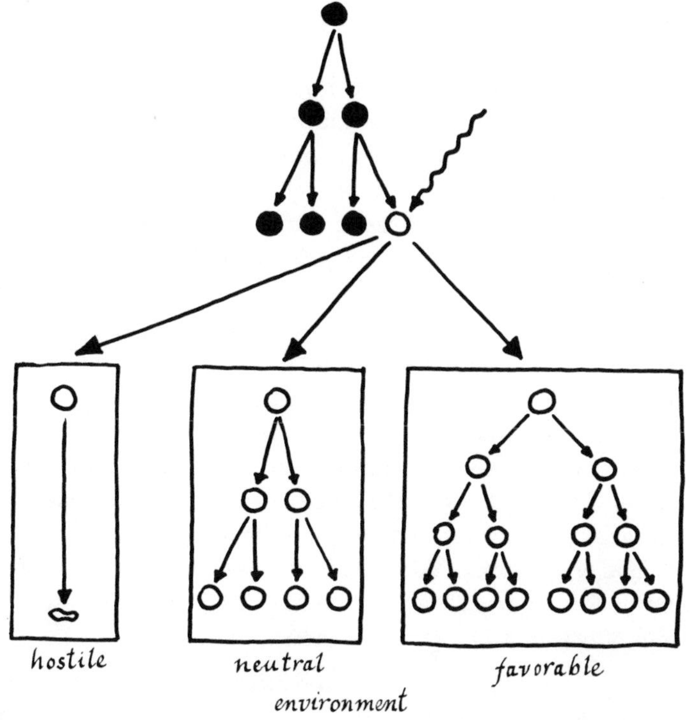

In principle, it is simple to measure the "success" of a change in DNA: count the number of individuals living a certain number of generations after the change occurred. If the total number of descendent individuals exceeds the total number of individuals living at the time of the change, the original change in DNA is beneficial, or successful; if the number of organisms has been reduced, the change is detrimental.

Similar considerations apply if a species or population of organisms is getting along happily and then its environment changes. Its ability to produce offspring will be (1) enhanced or (2) reduced. If the latter, eventual extinction can be avoided only if another change in DNA occurs, producing a variant that can reproduce *better* in the new environment.

In such simple relationships between change and selection lies the key to evolution. Altered DNA means changed protein; changed protein means changed organism. Changed organism finds itself in an environment not of its own choosing. Those organisms that have changes that allow them and their offspring to get along better will prosper; those organisms that have disadvantageous changes tend to die out. The natural environment *selects* – in favor of the better-endowed organisms; against the poorly endowed ones.

Remember that evolutionary success or failure is never measured in single organisms or immediately, as with our fat and thin friends in the North Atlantic. It is measured in large populations and over many generations.

The environment's impact on the ability of a species to produce offspring, then, is what it's all about. Rate of reproduction is the critical function.

Chance

Note again that chance plays the tune in evolution. How DNA will be altered by mutation is a matter of chance. What characteristics of a pair of parents will appear in offspring by sex-mixing of DNA is a matter of chance. The meeting of mating pairs is a chance occurrence. And what environment will be making the selection of changed organisms is in the hands of chance. Thus are the roots of all of life buried deep in chance.

Back to Bottles

Remember the beached bottles in Chapter I? Let's again imagine the bottles to be organisms. The chance event that changed the bottle, a bottle "mutation," if you will, was the inadvertent replacing of the cap. The relevant *environment* was the sea into which many bottles, capped and uncapped, were thrown. The sea made the *selection*: uncapped bottles were sent to the bottom; capped bottles floated on the surface until they were washed up on the shore as survivors.

We can now readily see that the bottle model for change and selection is badly flawed. That's because bottles can't reproduce their kind. It would have made a better story if a few *sexually active* bottles had been capped, dropped into the ocean along with a multitude of uncapped bottles, and thence found their way to shore while still reproductively vigorous. They would have then mated, producing children and grandchildren and so on to establish a thriving community of capped bottles on the beach.

Having endowed bottles with sexuality, we may as well carry bottle evolution one step farther. Suppose our beach over a period of time becomes stony, so that with each high tide the bottles get a drubbing. The many bottles made of relatively thin glass will not fare as well as those bottles made of thick glass. The thin-glass population, quite as able as the thick-glass population to resist drowning in the ocean, is now at a distinct disadvantage. Some will be able to reproduce some offspring, but many will not. The offspring, too, are at greater risk from stones, and soon succumb. In a few generations almost all the bottles on the beach would be of thick glass.

Moths

Some years ago certain white moths prospered around Birmingham, England.

They flourished on birch trees against the bark of which they were nearly invisible — especially to the birds that made a meal of them.

Over the years, Birmingham became heavily industrialized and the soot in the air gradually darkened the trees, and so exposed the moths. Against the now-darkened bark, they were sitting ducks for the birds. Consequently, the moth population, over many generations, decreased to near extinction.

During this period, an occasional dark gray moth was noted. This variety was well camouflaged against the gray bark of the trees.

Its numbers steadily increased, and these new moths eventually became plentiful in the region.

This story nicely reveals the interplay between organism and environment. Dark bark and predatory birds constituted an environment that selected *against* light moths, causing their population to diminish. A chance mutation occurring during this period produced a dark moth. While in earlier times, when the trees were light, such a mutation would have been deleterious, it was now beneficial, and such dark moths were able to mate and have progeny in peace. So they and their descendants prospered. The overall effect of a changed environment and a background of chance mutations was a complete change in the character of a population.

Bacteria, Too

Bacteria are excellent experimental models for studying evolutionary change — natural selection. They're a pure breed; all individuals in a population are identical because they're all derived from a single cell. And there's a new generation born every half hour, so you can follow populations over many generations in a reasonable time span.

Now, let's do to a pure population of bacteria in a glass flask in the laboratory something akin to what Birmingham did to its moths — create an unfavorable environment. We'll add a drop of the antibiotic streptomycin to the fluid in the flask.

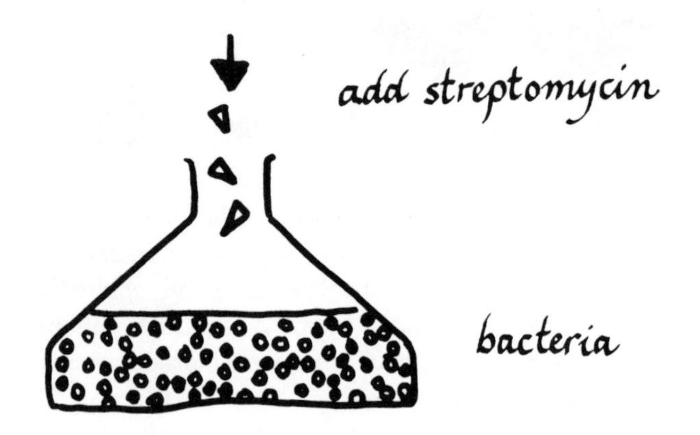

This is a catastrophe for these bacteria because they are killed by the drug. Growth stops fairly promptly, and cells begin to die. In a few hours *all* the cells seem dead.

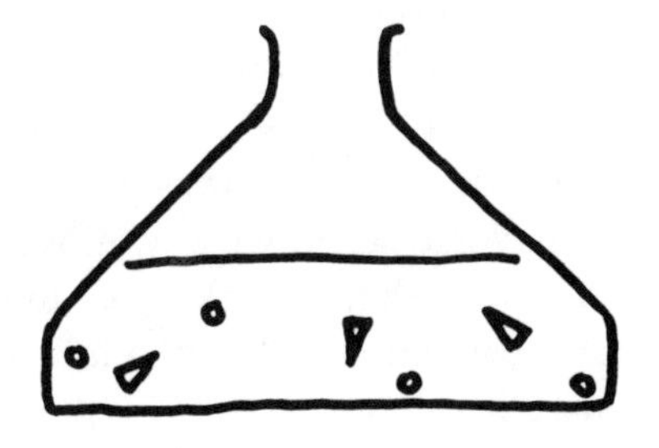

We can make a test to see if any living cells remain. We find there are a few living cells – say, fewer than ten – among the millions of dead ones. Furthermore, we can show that each of these rare living cells can multiply quite well in the *presence* of streptomycin!

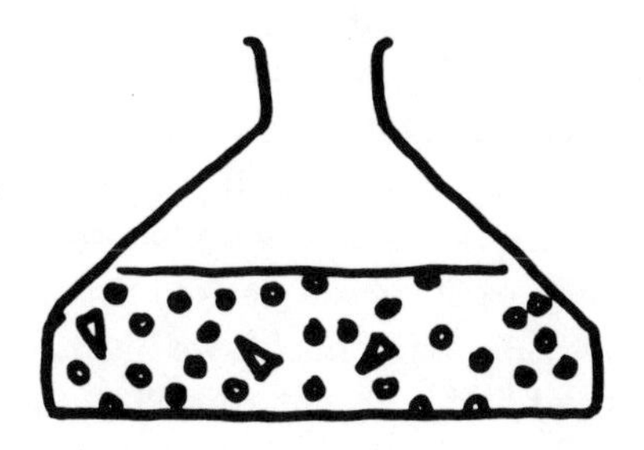

They're not in the least bothered by the drug. And this streptomycin-resistant trait breeds true; all future generations of those few survivors inherit resistance to the drug.

What's the explanation? In these extremely large populations of bacteria (many millions of cells) there's a good chance – perhaps 1 in 10 million – of a cell's undergoing a mutation that makes it resistant to streptomycin. Such a mutation will, of course, happen whether or not streptomycin is present; it is a purely chance change in DNA. If streptomycin were not present, we'd not know that the mutation had occurred. In streptomycin's presence, however, the resistant organisms are selected *for*: the organisms have a selective advantage. The streptomycin-resistant cells, therefore, keep dividing until they are the dominant population. The original bacteria die off because they are not equipped to survive in that particular environment. In all essentials, this story is identical to the moth saga.

Return to the Soup

In Chapter II we envisioned the progeny of the first cell on earth gluttonously consuming the rich soup in which it had its genesis. We should now qualify the picture to the extent that an organism's ability to eat — that is, ingest and generate energy from compounds such as sugars — depends on special enzymes. If the enzymes are not present in the cells, the sugar can't be used. An analogous situation would exist if we lacked essential digestive enzymes in our intestine; though we had access to certain foods, we'd not be able to get them into our bodies and burn them. Although the first cell on earth had the capability of using one or more sugarlike chemicals in the soup, it is reasonable to expect that it didn't have the ability to use all of the many chemicals available. So after it had consumed all the materials it could use, it would stop dividing, and remain in a state of "suspended animation." Bacteria today when starved of chemicals they need for food do just that — they simply stop and wait. With billions of cells waiting around in the soup, random mutations would be occurring over long periods of time. If some of these mutations gave an organism a new capability for using *another* chemical, it would be able to begin to multiply again. In this way the soup would eventually be consumed by an ever-increasing variety of organisms living in it.

Evolution in the Wild

The examples we've considered so far might be called domesticated evolution. There's a clear relationship between a *single* change in a population and selection for or against that change. The creatures we use in the lab are essentially pure strains; that is, they're genetically identical: every individual is the same, at least until a mutation is introduced.

In the natural world around us, although the same principles apply, the situation is more complicated. We seldom see pure strains in nature. In fact, what at first struck Darwin, and what should strike us as we look about us, is the great *variety* of living forms. But not just the variety of different types, or species, of creatures — the variety *within* species. Almost any trait or character you undertake to measure within a species will show great variability. We have only to look at the human species; though we're all human, we vary greatly one from another. And in other animals it's the same: thickness of fur, speed of running or climbing, length or sharpness of teeth, height, weight, strength, sense of sight, of smell, appeal to the opposite sex, all vary over a wide range.

If you measure a set of such traits in a pure-bred strain of mice, there would be no difference — all animals would be identical.

Variety is what allows evolution to work. Though

Darwin and Wallace didn't know the cause of variety — mutation and sex-mixing of DNA — they perceived its importance and developed their theory from it.

You must now extend your perception to grasp the idea that a given population at a given moment in the history of its development carries in its DNA a very large number of accumulated changes.

The population is, in fact, the repository of all past DNA changes *and* all past selections made by the environment. This accounts for the great variety of individuals within the population. It is upon this variety that selection works its way in the population's further development.

Let's take just one variable: running ability. Among a large herd of grazers on an open plain there will be a wide range of maximum speeds achievable on demand. With a good supply of lions lurking on the sidelines, the swifter runners will have a better chance of survival and of reproducing their kind. So, over many generations, given stability of the environment, the herd will become relatively enriched for fast runners; the average speed of the herd will increase.

You can yourself envision similar kinds of forces behind the emergence of other traits:

Environmental Change	*Selection Favors*
Forest to open plain	Good running legs
Plain to plain with predators	Better running legs
Forest floor to trees	Good branch-grasping arms
Land to air	Lighter bones, longer arms and feathers
Warm to cold	Fur, sweat pores
Meat fare to grass	Short, grazing teeth

Does Evolution Have a Purpose?

One of the problems in comprehending evolution derives from one's seeing changes that *seem* purposeful, when, in fact, the mechanism involves only chance events. For example, if animals in an environment containing plenty of other, smaller animals gradually develop meat-eating teeth, the change makes sense: those that will succeed must eat other animals, and carnivorous teeth permit them to do it. There seems to be purpose here, as though the environment were *directing* the animals to make the change for their own good. Indeed, it was just this kind of wishful thinking that swept T. D. Lysenko, Stalin, Khrushchev, and the whole Soviet Union into a scientific comic opera for nearly thirty years. Aside from the fact that there is no conceivable way an environment could instruct a population of animals to change, it just doesn't happen that way. Rather, a population has a great variety of tooth shapes and sizes by virtue of accumulated chance changes. At each turn of a generation's wheel, those animals with tooth structure more favorably disposed to killing animals and chewing their flesh will have a fractionally better chance of surviving and producing offspring. Gradually, with continued selection over many, many generations, a carnivorous species of animal will evolve. The process is entirely devoid of purpose.

The word "selection" is perhaps misleading, for it connotes purpose. The environment, of course, is entirely passive. The environment doesn't *cause* favorable or unfavorable changes to occur. The changes occur spontaneously (mutation and sex-mixing), and once made, may help an animal to get along better in the environment.

Look back at the moths for a moment. Among a large population of white moths, the occasional appearance of a gray moth is a purely random, chance event, unrelated to any "need" for gray moths. The event would occur just as frequently during the white-tree period as it did during the gray-tree period. The trees don't direct the occurrence of the mutation to grayness. However, if by chance the variant moth does appear during the dark-tree period, it is much more likely to survive and produce gray offspring.

The gray moth on the dark tree with birds is equivalent to the faster runner on the open plain with lions. If you understand these basic relationships, you have grasped the principles of evolution that Darwin and Wallace formulated in their great flash of insight after contemplating the variety of life on earth.

Mutation and Selection in Humans

Humans evolved from simpler forms by mutation and sexual selection just as did bacteria and moths. Even now, we can see certain aspects of the process at work. Some mutations in humans produce variation in the form of disease, caused by the alteration of a protein that has some important function in the body. The failure of the protein to function properly causes illness. There are large numbers of specific genetic diseases now known that have this cause; in each one a different protein, generally an enzyme, is not functioning properly. One example we've already discussed (see Chapter V) is sickle cell anemia. Here a mutational change in DNA leads to the production of altered hemoglobin molecules. The changed hemoglobin molecules, in turn, alter the shape of the red blood cells in which they are carried, causing sickness.

Now, there's not much good you can say about this disease. However, sickle cell anemia sufferers who reside in regions of Africa where malaria is endemic are protected against malaria by their disease! Malaria is caused by a parasite whose habit it is to bore into red blood cells and stir up trouble. The parasites don't like to tackle sickle-shaped red blood cells, preferring the cells of healthier victims instead.

This relationship of sickle cell anemia and malaria points up again the subtlety in the relationship between an altered organism — in this case, a human — and its environment. Although those who suffer from sickle cell anemia are obviously at an evolutionary disadvantage (because they are sick), in malaria country they may be at an advantage relative to those who will be made even more ill by malaria.

Variety of Species

Wherever we want to look, some kind of life will be found busily engaged in the job of surviving. Whether in a spoonful of soil or water, at any height or any depth, in hot springs and frozen tundra, in the ocean or in the air, in arid desert or steaming jungle, evolution seems to have found a place for every imaginable — and even every unimaginable — form of life. Every mode of sensing, eating, moving, communicating, loving, fighting, protecting, and reproducing is being put to use. And what we see on earth today is only a minuscule fraction of the varieties of living creatures that have lived and vanished forever. The dinosaurs are monuments to unnumbered birth-success-defeat-extinction cycles that have run their course over thousands of millions of years.

Can change and selection explain all that diversity and complexity? Given our ignorance of the details of how it all came about, we can say only that in principle the interplay between change and selection *can* account for the observed ever-expanding complexity. It is a sufficient explanation. Changes that give organisms *added* capabilities increase their chance of survival. Given enough time, everything will be tried.

But of one thing we can be sure. If we'd lived 2 or 3 billion years ago and tried to look forward, we could not possibly have predicted what actually happened. No one could have anticipated humans, or any other living form, for that matter. Why? Because every step in evolution is a chance — and therefore unpredictable — event. All living creatures, humans included, are products of an enormously large series of chance events. It may be said that in the particular form in which we humans find ourselves today, we are incredibly improbable! Another way of saying this is that if evolution started all over again on the same earth and under the same conditions, the chance of producing humans again would be infinitesimally small. It may also be said, incidentally, on the basis of the same reasoning, that the probability of there existing anywhere in the universe creatures resembling us is vanishingly small. The likelihood of life is great, but the likelihood of life familiar to us is very small.

We conclude that change and selection are *sufficient* to explain the human presence. And science always likes sufficient — yes, simple — explanations.

CHAPTER VII

Embryogenesis

The most fascinating and challenging of all problems in biology is embryogenesis: creation of the embryo. Embryogenesis is concerned with the steps involved in transforming a single cell, the fertilized egg, to the complex, many-celled organism it is destined to be. That destiny is written in DNA, which provides all the instructions for this exquisitely orchestrated series of developmental events. I'll tell you right now that we don't understand this wonderful process, but we can at least explore around its perimeter.

Cells Stick Together and Specialize

A fertilized egg, like other simpler one-cell creatures, begins its life by dividing to make two cells; these two divide to make four and so on. And we expect, from observing one-cell creatures, that after each division the cells will separate. But the descendants of a fertilized egg *don't separate.* They stick together, as though they knew they were participants in a communal enterprise.

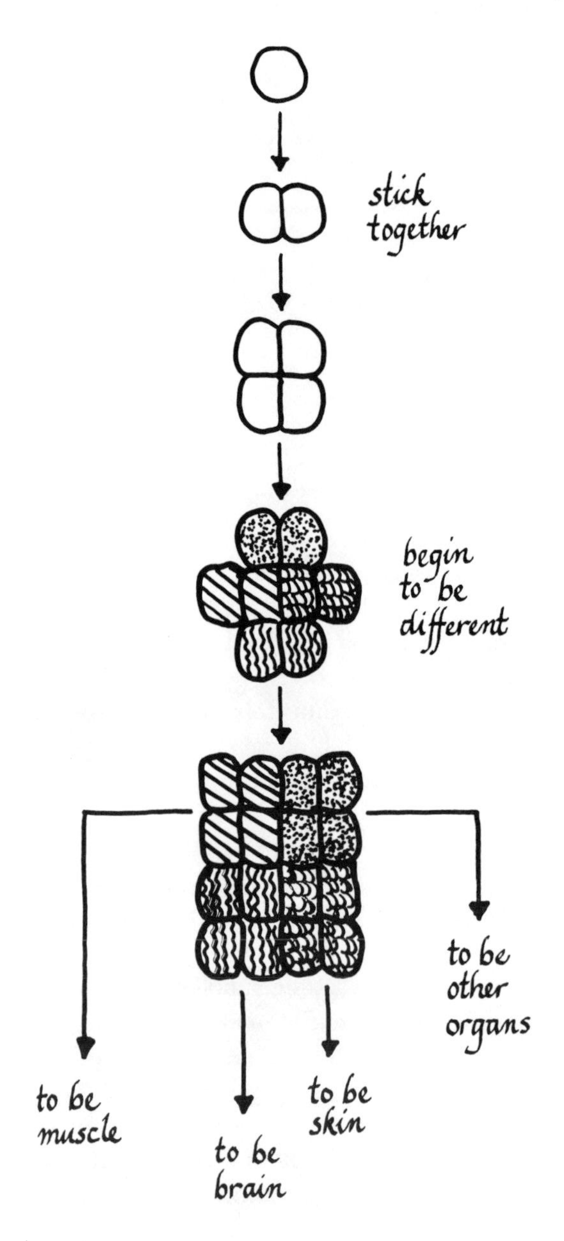
stick together
begin to be different
to be other organs
to be muscle
to be brain
to be skin

Another thing soon becomes apparent: cells are segregating into groups that look and act differently. Groups of cells are, in fact, becoming specialized — each group involved in doing a limited number of special tasks. The specialization process is irreversible; there's no turning back.

The two features of early embryogenesis — *cell stickiness* and *cell specialization* — appear to be at the very roots of the development process.

The Origin of Difference

Up until now we have been learning about laws, applicable to all living creatures, that determine how organisms gradually become different over long periods of time. All living creatures store their information in DNA, transcribe the DNA into messenger RNA, translate the messenger RNA into protein. Furthermore, alteration of DNA by mutation or sex-mixing causes proteins to be altered permanently, thereby gradually introducing cumulative differences among organisms, and finally new species.

In some ways embryogenesis resembles evolution in microcosm and over a short time span. As we watch an animal embryo proceed through the various stages of development, it resembles a fish before it resembles the adult creature it is destined to be. The fish resemblance is more than superficial; the early embryo has real gills for underwater breathing. Since the embryo doesn't need gills — it gets its oxygen and food from the mother via the umbilical cord — there appears to be no obvious explanation for the embryo's apparently repeating a stage in evolutionary development.

But when we ask: How does *difference* arise during embryogenesis? — How do cells make the decision to become skin cells, muscle cells, nerve cells, and so on? — Nature responds with a blank stare. She's permitted us to learn a great deal about the universal mechanisms of information-processing in cells, but when it comes to those things that make cells different from one another, we have been kept in ignorance. Some scientists believe we'll need totally new concepts and methods to plumb the depth of embryogenesis. I suspect this is not so. It's likely that the things that make cells become different are more complicated than the things we've discovered so far.

Medicine's Interest in Embryogenesis

An understanding of embryogenesis is important to medical science. It's not just that the transformation of a single cell to a complete individual whets the medical scientist's curiosity like no other happening. It also has to do with medicine's search for better control of problems related to pregnancy, birth control, infant mortality, congenital disease, genetic disease, and cancer. Scientists have a presentiment that an understanding of embryogenesis will shed light on a large number of these baffling medical problems.

More about Stickiness

I mentioned that after the fertilized egg begins to divide, the cells stick together. What makes them stick? One thinks of glue, but it's not really a glue. It's more as though the cells had burrs on their surfaces — tiny hooks that snag the hooks of other cells. The cell's DNA has, in fact, instructed the cell's protein-building machinery to make certain specific proteins that migrate to the outside of the cell and there act like the hook of a burr. As cells become specialized to be the different parts of the body, their surface protein hooks also become specialized. By this means, cells of a particular type recognize one another.

Energy for Embryogenesis

You are certainly now properly sensitized to the need for energy in all construction processes. Sugar must be supplied to a developing embryo so that its cells can burn to produce ATP. In fish, reptiles, birds, and other creatures, where embryos grow inside an egg, the yolk of the egg provides food for the embryo. In animals that grow in their mother's uterus, a different device is used. Between the inside wall of the mother's uterus and the embryo, the placenta grows apace with the embryo. The placenta is where the mother's blood and the developing embryo's blood meet. The bloods, mingling, bring the food eaten by the mother to the embryo. This is how energy for the construction project is made available.

The Same Information Is Distributed to Every Cell

The fertilized egg starts off with the full amount of DNA from mother and father, and as it divides, each successive generation of cells receives the same full amount — right on up to the adult. Thus, a human being composed of 60 trillion cells contains 60 trillion identical copies of DNA! So every cell of the body contains exactly the same information (except reproductive cells, which contain half the DNA of other cells).

Controlling Gene Expression

The secret of embryogenesis, when we learn it, will be found in the way cells control the expression of the genes of DNA. All the information is there, waiting to create an adult. If we could peer deep down inside each cell of the developing embryo we would observe certain things happening. Enzymes would begin to copy some of the genes of the DNA of the fertilized egg into messenger RNAs. These messenger RNAs would go to the ribosomes, which came along with the egg to get things started, and there begin the synthesis of the requisite proteins. After the fertilized egg contained all the prescribed proteins, including some more ribosomes – and had doubled its DNA – it would divide. The resulting pair of cells would now contain a new full measure of DNA, new ribosomes, new everything. Indeed, it would be an exact copy of the cell from which it came. The process of protein synthesis and new-cell construction would repeat itself, resulting in four cells. And so again for eight cells.

Thus far the process is pretty much the same as what goes on in dividing bacteria. Each generation is an identical repeat of the previous generation: DNA→messenger RNAs→proteins→cell division. However, when specialization gets started, something new must happen. If the progeny of one group of cells is going to become skin, another muscle, another brain, and so on, DNA must provide the necessary guidance. It must determine not only an ever-widening difference among cells, but also *when* the difference will appear.

Yet every single cell in this developing cell community contains exactly the same measure of DNA. How then can the cells become different?

First, remember that the character of a particular cell, be it skin cell, muscle cell, or brain cell, is determined by the proteins it makes. Skin cells, for example, are rich in a special protein called *keratin* — a protein that gives skin its special ability to protect us. Muscle cells abound in a protein called *myosin*, which has the special ability to interact with a partner protein and so change its length and permit muscle fibers to contract. Brain cells contain proteins that assist in transmitting electrical impulses. And the cells of each of the other specialized tissues produce their own unique proteins, which in turn determine the character of the cells.

So some cells will start producing keratin to realize their destiny as skin cells; others will start producing myosin, to become muscle cells. Well, *all* DNA has a gene for keratin and another gene for myosin.

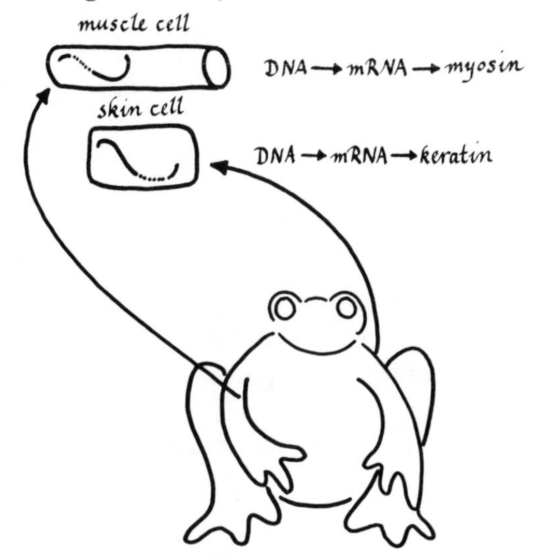

The genes are there. It seems, then, that the gene for keratin must be *expressed* in skin cells, while the gene for myosin must be *suppressed.* And, on the other hand, the gene for myosin must be *expressed* in muscle cells and the gene for keratin must be *suppressed.* So the keratin gene in skin cells is read out as a keratin messenger RNA. It goes to a ribosome and is translated into the protein keratin. The cell in which all this is happening then becomes a skin cell.

DNA must be able to express its genes and suppress its genes in a programmed time sequence during embryo development. It takes hundreds of proteins to make a cell of a particular type, so in these cells many genes are expressed while many more genes — those for proteins of other types of cells — are suppressed.

A remarkable state of affairs indeed! DNA contains all the genes but also has information as to when these genes should be turned off.

The Nature of the On-Off Gene Switch

Here, then, is a point of focus, a phenomenon about which one can ask a sharp question. How can genes be turned on and off?

As we have seen before, the clearest answers come from the simple systems. Observe again the behavior of the lowly bacterium. If you suspend some cells in a fresh growth solution and add a sugar, let's say glucose, the cells begin to divide and the number of cells increases rapidly. This goes on until the glucose is used up. Growth then stops.

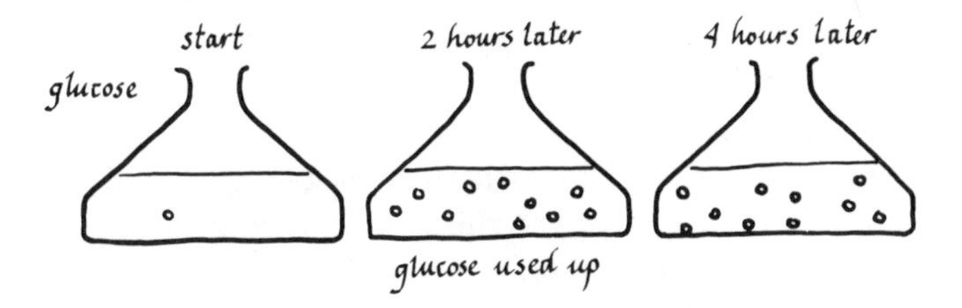

Repeat the observation with an identical group of cells, but give them a different sugar — say, galactose — and the number of cells increases, but at a slower rate than with glucose, stopping when the galactose is all consumed.

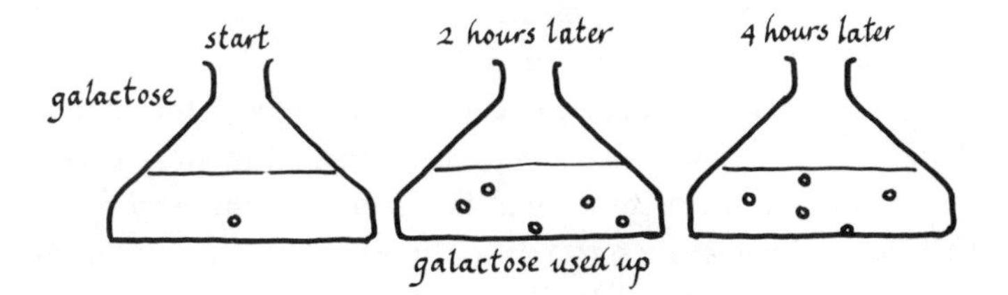

We conclude that glucose is a better food than galactose because it can be consumed more rapidly. But *both* sugars are used up by the bacteria. They don't waste either.

Now repeat the experiment using *both* glucose and galactose. An interesting thing happens. The population increases rapidly until glucose is all used up. Growth then stops for about twenty minutes. And then growth starts again and continues until galactose is used up.

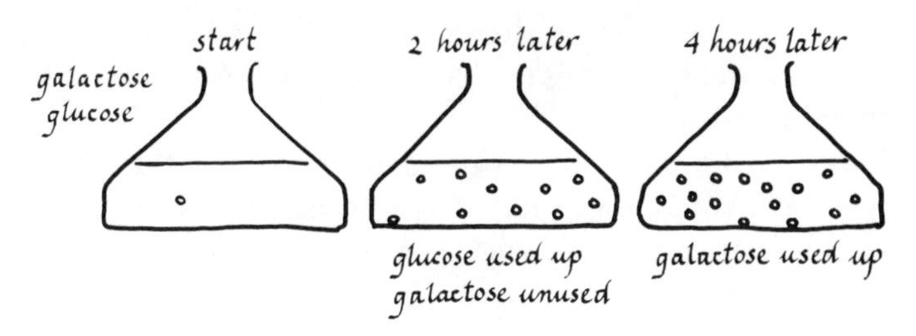

Obviously the cells prefer glucose and then, after a twenty-minute lag, they *acquire the ability* to use galactose — after glucose is gone.

What's this got to do with turning genes on and off? Well, it was the analysis of this simple system that led in the late 1950s to a brilliant new insight into the control of gene expression by the French scientists François Jacob and Jacques Monod. The mechanism is now proven to be operative in bacteria; it probably functions in more complex organisms, such as ourselves, but we don't yet know for certain.

Now what was going on in these bacteria as they dealt with the unaccustomed surfeit of sugar? The bacterial cells obviously contained the "machinery" for using glucose — they started eating it as soon as they were given it. This machinery consists of two proteins: an enzyme to facilitate the sugar's entry into the cell, and an enzyme to digest it once it's inside. Two enzymes; two genes. They apparently did not have the corresponding machinery to handle galactose, at least when growth on the two sugars started. Yet they *acquired* the machinery to use galactose when glucose was used up. The absence of glucose, then, triggered the development of the machinery to use galactose. *Glucose was preventing, or repressing, the expression of the genes that controlled the enzymes to use galactose.* When glucose was gone, the repressive effect of glucose disappeared so that the genes for galactose could begin to make messenger RNAs and so be translated into protein.

Think what this means to the bacterium. It eats the best food available, and that food, inside the bacterium, arranges it so that no energy will be wasted in making enzymes to use another food. When the good food is gone, and the poorer food is the only one available, then the bacterium can go ahead and make the enzymes for using the latter.

Bacteria Don't Make Things They're Given

If you were growing vegetables in your own garden, of necessity, for your own consumption, and someone agreed to supply you regularly with those same vegetables, you'd probably stop producing your own. Bacteria do something similar. They're able to make their own amino acids — the twenty essential links in protein chains. Without amino acids they could not, of course, build proteins, and so would stop multiplying. Now, if we *give* bacteria amino acids already made (simply by adding them to the solution in which the bacteria are living) they *stop making* their own amino acids. The gift of amino acids makes it unnecessary for the cells to expend energy in making their own. And considerable energy is involved. The making of each of the twenty amino acids requires several enzymes. When each enzyme is made, a gene must be turned on, messenger RNA made and transported to ribosomes where the enzyme protein is made. Shutting off the gene thus means a considerable saving in construction energy. Energy conservation, as with all living cells, is vital to bacteria for their survival.

Scheme for Control of Gene Expression

So here's the general picture of gene expression that has emerged from studies on bacteria.

1. Genes *can* be switched on or off. It's done by protein molecules called *repressors.*

2. Repressors affix themselves to one end of a gene so that the enzyme that would transcribe it into a messenger RNA is prohibited from doing so.

3. This means that the protein the gene is responsible for can't be made.

4. Repressors are *released* from the DNA by two kinds of things:

a. by the *absence* of a sugar like glucose (that is, glucose helps the repressor attach to the gene);

b. by the *presence* of an amino acid.

Now we can see the explanation of the glucose-galactose experiment we discussed earlier. As long as glucose is available to the bacteria, they will eat it and it will assist the repressor of the galactose genes to keep them shut off. When glucose is gone, the galactose-gene repressors can't function, so the enzymes are made and galactose can be used. In like manner, when bacteria are given amino acids, they can make the amino acids assist the repressors of all the genes involved in making amino acids, thereby shutting them off.

This very handsome system for regulating things in bacteria also appears to be operative in higher forms of life, including humans, and is certainly an important way of controlling the expression of genes.

But Humans Aren't Bacteria

Now, there's a very significant difference between the ways of bacterial cells and the ways of more complex and specialized cells of organisms like us. The bacterial cells live a life of quick response, flexibility, rapid adjustment to drastic changes in their environment. It's pretty much a jungle warfare existence: every bacterium for itself.

Specialized cells, on the other hand, are permanently committed to a particular lifestyle. Skin cells remain skin cells, muscle cells remain muscle cells, brain cells remain brain cells throughout the life of an organism. In each type of cell the few genes that determine skinness, muscleness, brainness are turned on, and all other genes — say, those for liverness, boneness, kidneyness — are turned off. And this is the way it stays for a lifetime. Bacteria, therefore, need devices for turning genes on and off rapidly and easily. Specialized cells must have most genes turned off permanently, a few on permanently. So the easy on-and-off switching mechanism of bacteria may well not be similar to that which is operative in specialized cells. At the moment, however, the bacterial system offers the best *model* we have, and in theory it would not be hard to make it work as a permanent on-and-off control.

The Genesis of Form

We've been looking at the bedrock problem of gene expression in embryogenesis. But the aspect that more immediately strikes the eye is the genesis of *form*: the artistry of the sculpturing process, the incredible architectural achievement of the egg-to-baby transformation. For example, all of the special tissues and organs of which we're made are "hung" on a skeleton. Bone develops within the embryo, along with every other structure. Beginning with some ordinary-looking cells, a new tissue appears in which calcium can be deposited to make a hard structure. Hard and extraordinarily strong, ingeniously made to carry the weight of an organism for a lifetime, and capable of growth and repair when fractured. How do such structural shaping processes come about? This is a tough problem to get a grip on and again we have recourse to a model system.

Bacteria, like humans are susceptible to infection by viruses. Each bacterial virus (called a *bacteriophage*, meaning bacteria eater) has a boxlike "head," into which its DNA is packed, and a "tail," which acts like the needle of a syringe. At the point of the needle there are spiderlike legs, which grasp the bacterium's surface.

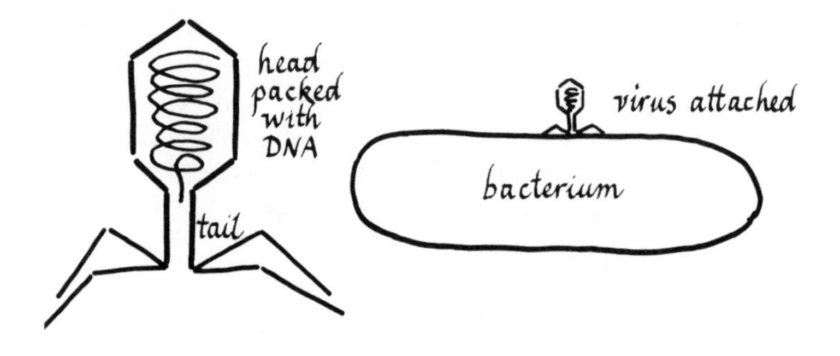

The virus then *injects* its DNA through its tail into the bacterium as though the virus were nothing but a syringe — which is exactly what it is. Once inside the bacterium, the virus's DNA immediately takes command. A signal goes to the protein-construction apparatus of the bacterium, indicating that henceforth no more bacterial protein will be made. The ribosomes and transfer RNA apparatus is quickly appropriated by messenger RNAs produced from the virus's own DNA, and soon the bacterial factory is making virus protein parts — new heads, tails, and tail fibers. The whole plant is commandeered by the virus's DNA.

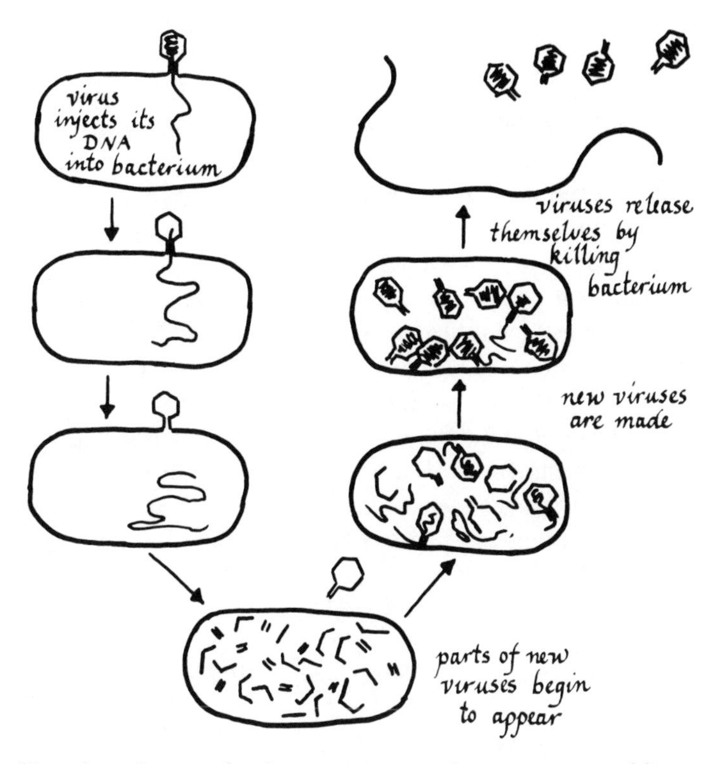

Shortly afterward, virus parts can be seen assembling inside the bacterium, newly formed virus DNA is packed into the heads, and complete viruses materialize. About 100 viruses accumulate in each bacterial cell, pretty much filling the bacterium's insides. And then — the coup de grâce — the viruses secrete an enzyme that destroys the bacterium's membrane, killing the bacterium and allowing the new viruses to escape. This whole outrageous subversion takes less than half an hour!

We can see in this phenomenon a simple model system for the genesis of form. The different parts of the virus are put together like a small building under the instructions of its pure DNA, using the commandeered factory. It has been shown to be an assembly process that has a carefully programmed time sequence, such that the genes controlling the manufacture of the different parts of the virus are activated in sequence. There's a strong indication that if the right parts are made in the right sequence, the definitive form will evolve spontaneously.

How useful the model will be in shedding light on the much more complicated phenomena of true embryogenesis remains to be learned. The usefulness of the model resides in the facts that we have a rather complete knowledge of the gene composition of the virus (a much simpler organism than a bacterium), we can control and manipulate the sequence of events, and we can follow the genesis of a not-too-complex three-dimensional form with ease in the electron microscope.

Starting and Stopping Cell Division

An embryo is a mass of rapidly dividing cells. This vigorous growth activity continues after birth and through childhood at a slowly decelerating rate, until adulthood is reached. Then cell division stops. Throughout an organism's whole body, the cells of every organ, every tissue, participate in a carefully coordinated termination of growth. How do the cells know when to stop growing? What tells them that the organs of which they're a part are now just the right size?

This phenomenon can be observed in the behavior of normal cells outside the body. They will grow over the glass surface of a dish, staying only one layer deep, always in contact with their immediate neighbors; and when the outriding cells reach the edge of the dish *all* the cells stop growing.

What's the nature of the division-stopping signal(s)? We don't know the answer and we continue to search for it. There is a challenging model system that should yield answers to at least part of the puzzle. I have a special fondness for it, having spent many years bewitched by its pertinence.

Regeneration

If I cut off a tadpole's tail and then drop the tadpole back into water, the stump heals quickly, and over the course of the next three weeks a truly remarkable thing happens: the stump produces a completely new tail.

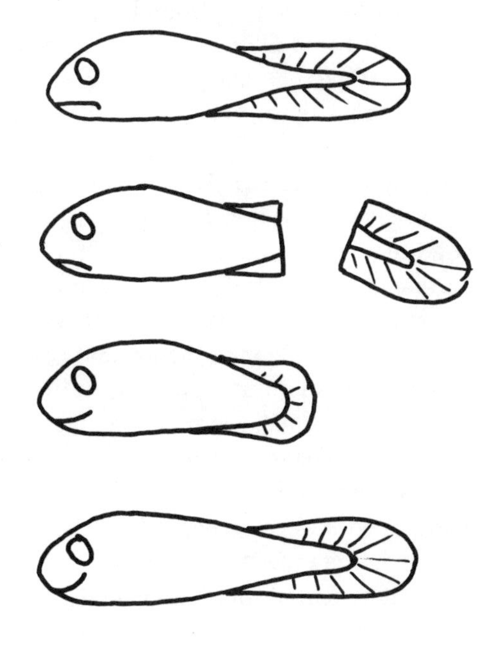

Complete and perfect. A salamander similarly restores a leg if I amputate it. Starfish and lobsters do the same. The phenomenon is called *regeneration*, and it happens to us, too. We can't restore legs and arms, but if, for instance, our liver is damaged in an accident, necessitating surgical removal of part of it, the liver quickly — within a few days — regenerates to its original size. This particular situation can be simulated in the laboratory. I can surgically remove two thirds of a rat's liver. The rat recovers from anesthesia in a few minutes, begins eating in a few hours, and three days later the missing two thirds of the liver is back again, normal and healthy, doing all the things a liver should do.

Two dramatic things happen in all these cases. First, the removal of a part of the animal inaugurates a very rapid rate of cell division at the amputation site, where all was quiet before. Second, when the part has been restored, cell division ceases. The incredible thing is that the cells in the part that was left behind "knew" of the need both to start dividing and to stop when the job was done!

What is it inside these cells that tells them to start dividing, and to stop when they have divided enough to restore the missing organ? I once tried to get at the answer by breaking apart cells from regenerating liver and mixing them with broken cells from normal nondividing liver. I thought that if there was a chemical "signal" in regenerating liver cells, telling them to make more of themselves, it might influence the normal cells and make them construct protein more rapidly. If, on the other hand, the normal cells contained a chemical signal that could tell regenerating liver cells to slow down, I could pick that up, too. A good idea, a good use of a valuable model system — but the experiments were inconclusive. The system is yet too complicated, and eludes our grasp.

In our story of repeated successes in elucidating the laws

of life, an experimental failure seems out of place. On the contrary, it enhances the truthfulness of our tale. For by far the greater number of experiments scientists do are failures. We learn from our failures, and design better experiments that may eventually give us a new insight.

A colleague of mine, Dr. Nancy Bucher, has probably contributed more to our knowledge of regeneration than any other scientist. Some of her important work has involved making Siamese twins of rats. She sewed rats together side by side until they established a good shared circulation; until blood flowed easily between the two. She then removed two thirds of the liver of *one* of the rats, and, as its liver regenerated, watched to see if the liver of the *other* rat began to grow. It did! This meant that the regenerating liver caused something to go into the blood stream, where it could circulate to the other normal liver and make it start to grow. She and many other scientists have tried to learn what this substance might be, but have so far been unsuccessful.

Ignorance of Embryogenesis Exceeds Knowledge

So, to recapitulate: we've considered some interesting things about embryogenesis: the ability of dividing cells to acquire the stickiness needed to remain together; the advent of the specialization necessary to make a complex organism; the genesis of form and shape; and, finally, the stop signal — needed to bring the long process of embryogenesis, childhood, and adolescence to completion. These are only a few of the highlights of an incredibly complex process. Our ignorance still far exceeds our knowledge. This is not surprising: embryogenesis seems to be a problem that demands that we use every skill we have. It lies at the very roots of the science of biology. But it is particularly exciting and provocative partly because there seems to be nothing about it too difficult to solve. I believe we'll soon understand embryogenesis as we now understand the universal laws of life we've discussed in earlier chapters.

The nature of the problems presented by embryogenesis are very similar to the problems presented by cancer. Indeed, some researchers feel that an explanation of cancer will require an understanding of embryogenesis. Cancer, in some senses, seems to represent a loss of that superb quality of *control* one sees in embryogenesis. The truant behavior of cancer cells, for example, might be related to a loss of cell stickiness. We must now examine these matters more closely.

CHAPTER VIII

Cancer

The trouble begins when something happens inside a normal cell that changes it into a cancer cell. It is this single-cell origin of cancer that makes a chapter on cancer appropriate in a book about the principles of life. Indeed, it turns out that those who seek an explanation for cancer need to be concerned with many of the matters we've already discussed in this book.

Beryllium and Cancer

The metal *beryllium* produces extremely malignant bone cancers in rabbits. This was discovered, about the time I began my research career, by some scientists who were concerned about the death of workers in the fluorescent lamp industry. Their lungs were being destroyed by exposure to the materials used in the lamp tubes. The researchers injected the phosphorescent material, which contained beryllium, into animals, and the bone cancers appeared several months later. The cancers grew rapidly, spread to other parts of the body, and killed the animals in a matter of weeks. (Beryllium never produces cancers in humans, but does damage their lungs, and is no longer used in the manufacture of fluorescent lamps.)

Fresh out of medical school, I undertook the task of finding out more about how beryllium caused cancer of bone. First, I searched the library to discover what other scientists had and had not done to learn about beryllium's action on living systems. Very little had been done. One interesting thing was known, however. Beryllium, in extremely small amounts, stops the action in the body of an important enzyme called *phosphatase.* This enzyme is particularly important to bone because it helps to deposit calcium phosphate, the material that makes bone hard. Another fact was known: the enzyme phosphatase normally needs the metal *magnesium* to operate. Beryllium is a very close relative of magnesium in its atomic structure. So the question arose: Was beryllium poisoning phosphatase by competing with magnesium? The answer turned out to be yes. Beryllium enters the enzyme, pushes out the magnesium, and so paralyzes the enzyme.

Beryllium and Growth

This was where I came in. In pondering the problem, I came up with what seemed to me to be a good idea: Why not work with a simple model system that clearly required magnesium for *growth*? If beryllium could be shown to have some effect on growth, and if this could be related to the system's need for magnesium, I might be a little nearer an understanding of how beryllium influences cell growth. The simple model system I chose to work with was the growth of plants. The chlorophyll molecules in plants all contain magnesium. Chlorophyll, just like some enzymes, can't function properly without magnesium. Perhaps beryllium could be shown to push magnesium out of chlorophyll, thereby altering the growth of plants.

I began my work with tomato plants, grown in a greenhouse in bottles that contained a solution with all the necessary plant nutrients, including a normal amount of magnesium. These plants flourished for several weeks.

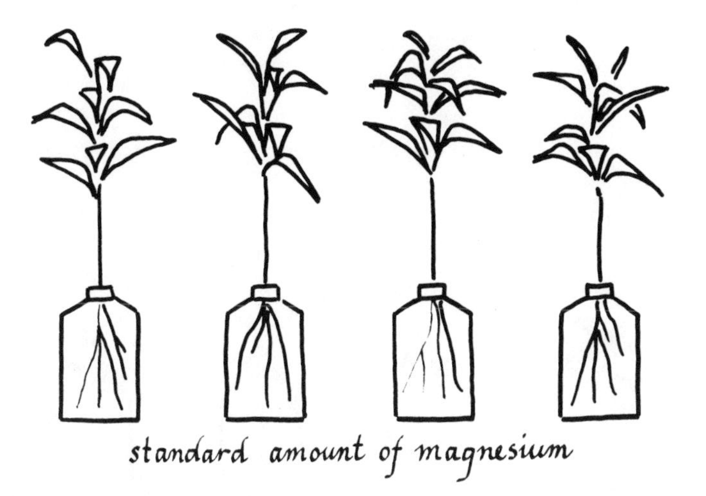

Another group of plants was grown in a similar environment, but with beryllium also present. Their growth was identical to that of the group without beryllium.

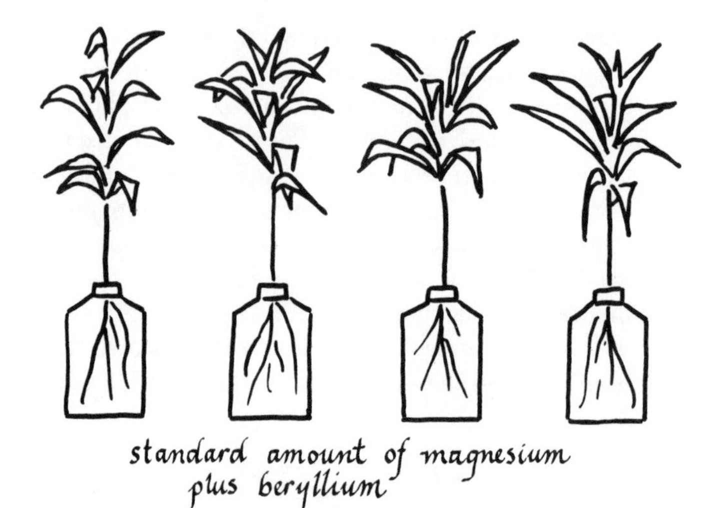

So I concluded that beryllium didn't cause any problem when magnesium was plentiful.

A third group of plants was grown with half the standard amount of magnesium. These plants grew for about a week, then turned yellow, wilted, and died.

This was the expected effect of magnesium deficiency: the half-amount of magnesium could not sustain them. A fourth group of plants was deprived of magnesium exactly like the third, *but* received the same amount of beryllium as the second group had received. The outcome was dramatic and gratifying. These plants flourished and in every way seemed to be doing as well as the first and second groups.

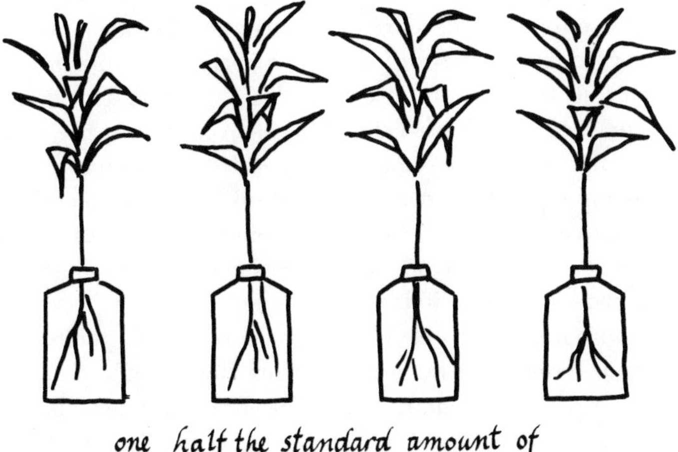

one half the standard amount of magnesium plus beryllium

My conclusion was inescapable: beryllium could replace at least half of a plant's magnesium requirement, allowing plants doomed to die of magnesium deprivation to grow quite normally.

So far, so good. The next step was to see if some beryllium had got into the chlorophyll molecules and pushed some magnesium out. I analyzed the chlorophyll from all four groups of plants and, to my chagrin, found that it had all of its magnesium, and not a trace of beryllium. It was discouraging, but I couldn't argue with the fact.

Yet there could be no doubt that beryllium was doing something magnesium normally does — to something in the plants. I repeated all the experiments, using cultures of tiny one-celled green algae, and got exactly the same results.

Why do I tell this story? First, it's an intriguing problem in cancer research; an interesting model system; a challenging unsolved mystery. Second, it illustrates the way events often develop in science: an idea leads to an expected dramatic result, but the anticipated explanation turns out to be wrong. A better idea is needed. Third, my experience with this beryllium problem convinced me, at a critical career-choice moment, that I could generate an idea and test it by experiment, and so could make a career in science. That my theory was wrong did not trouble me — most ideas are wrong, and one is very lucky to have a few good ones in a lifetime.

I pursued the problem for two more years, published some interesting discoveries showing that beryllium competes with magnesium in magnesium-dependent plant enzymes. But neither I nor anyone since has been able to explain the effects of this simple metal on rabbit bone, to say nothing of plant growth. An important unsolved problem awaits an ingenious solution. Certainly the production of cancer by a simple metal poses a challenge worthy of the most able scientist.

What Is Cancer?

Let's get a very clear picture of what cancer is. As a first approximation, we can say that cancer is an inherited abnormal behavior of cells. This particular abnormal behavior may start in any cell in any part of the body at any time. There are two main features of cancer-cell behavior. (1) Cancer cells multiply relatively more rapidly than their normal neighboring cells. As we learned in the last chapter, normal cells have a period of growth but then their growth stops. Regenerating liver cells grow too, but their growth stops when the original mass of liver is restored. Cancer cells never stop dividing as long as they have a supply of food. (2) Cancer cells are altered in their usual relationships to surrounding cells such that they become relatively more independent, asocial, unneighborly. You will remember that cell stickiness – the tendency of dividing cells to stick together in a neighborly fashion by virtue of special proteins on their surfaces – was an important feature of embryogenesis. The loss of this fundamental property of normal cells seems to be a very important feature of the switch to malignancy.

The combination of these two features – increased rate of cell division (growth) and loss of cell stickiness – is deadly. It means a new and "strange" tissue is growing, unregulated, in the body, and rapidly spreading away from its point of origin. Eventually cells may metastasize; that is, pass through the bloodstream to other parts of the body, starting new colonies of growth there. And in time these dividing, restless cells kill the body in which they were born.

Cancer Outside the Body

When medical scientists can get a problem out of the body and into a simple glass dish (the modern equivalent of the test tube), they become optimistic about solving it. For it means that their ideas can be put to critical experimental test in a controllable, manipulable system. Cancer is, as we've said, an affliction of cells. Cells can be removed from the body and studied in the laboratory in glass dishes. There are few human diseases that can be studied so easily.

Let's observe the behavior of normal cells and cancer

cells in glass dishes. First, we'll put a few normal body cells in the middle of a dish and cover them with a fluid to nourish them.

sideview of glass dish with single normal cell placed in the middle

Over a period of several days the cells divide repeatedly, always staying in contact with the glass and with one another.

3 days later cells are dividing, covering glass one layer deep

When the cells reach the edges of the dish they stop dividing.

cells have reached edge of dish, further growth has stopped

Thenceforth the cells remain in stable, neighborly contact, one layer deep. If we scrape some cells off the glass, the cells near the "wound" begin to divide, and soon fill the gap. When the gap is again covered with one layer of cells, cell division again ceases.

You will note that this kind of behavior is a simpler, but fundamentally similar, version of the behavior of regenerating cells: division continues vigorously until some preset limit, the original organ size, is reached. Both systems demonstrate convincingly that normal cells "know" when to stop growing.

Now we'll watch the behavior of cancer cells. We put a few of them in the middle of a dish and watch.

Cancer cell placed in middle of glass dish

They divide and cover the glass, looking not too unlike normal cells.

a few days later, cells are dividing

cells continue to divide

When they reach the edge of the dish, however, any similarity to normal cells ceases.

cells continue to divide

They keep right on dividing so that more and more cells pile up, one on top of another, in disorderly fashion. The cells seem to have "forgotten" how to stop growing. The only thing that now impedes continued growth is the availability of nutrients. Cancer cells have taken on a property that no other cells have: the ability to grow indefinitely, as though they were immortal.

Indeed, some cancer cells have been growing outside their victims' bodies for a very long time. The most famous instance is that of Henrietta Lacks, who, in 1951, was operated on for a cancer of the cervix. She later died of this cancer, but some of her cancer's cells had been put in glass containers and given nutrients, and they continued to divide. And, known as HeLa cells, they are alive and still dividing today! They are the cells most commonly used in cancer research.

As in the body, so in glass dishes. The restraints that cells normally apply to their neighbors have been dropped. The inhibition of cell division that comes about when cells have filled a certain space, reached a certain predetermined total mass, is lost in cancer.

But there is something more that can be done in glass dishes: normal cells can be induced to turn cancerous. By adding agents that cause cancer in animals, notably certain cancer-causing viruses, one can transform cells to malignancy outside the body. This is very exciting to the scientist, for it means that each step in the causation process can be followed under controlled conditions in the laboratory, outside the body.

Cancer Blood Supply

It may be argued that cells growing on glass are not typical and that they would be more representative of cancer cells growing in the body if they were grown in three dimensions in a setting similar to soft tissue. When this is done, cancer cells divide until they form a tiny ball just barely visible to the unaided eye. They can grow no further, apparently because they have inadequate access to their food supply. If there are blood-vessel cells nearby, the little ball of cancer cells will stimulate them to produce new blood vessels. The blood vessels grow into the cancer mass, and soon the cancer cells begin dividing again. And, as blood vessels carry nutrients to the cancer, and grow in parallel with the cancer, the cancer mass can grow quite large. This is the way cancer behaves in the body: it can't grow without its own blood supply. This important research, conducted by Judah Folkman, also shows that the cancer cells secrete something that causes blood-vessel growth. The search for this something is going on now. If we knew what it was we might be able to counteract it and kill cancers by starvation.

Is Cancer Due to a Mutation?

What could conceivably cause a cell to take on these outrageous properties? Of course, we don't know. What triggers the transformation is the big question. But there are several things about the way cancer gets started in the body that suggest a mutation; that is, an alteration in the DNA of a single cell.

1. Cancer seems always to start as a sudden change in a single cell.
2. Once a cell has become malignant, *all* of its descendants are malignant, too. That is to say, the malignant trait or character breeds true.
3. The cancer cells appear to have acquired a selective advantage over the normal cell from which they are derived.
4. Most of the things that induce cancer, like chemicals, x-rays, and ultraviolet radiation, also produce mutations.

So it seems that a possible immediate cause of cancer may be a change in DNA, a mutation.

Viruses and Cancer

Certain kinds of viruses can cause cancer. We're going to see that this fact has an interesting relationship to what we've said about mutations.

Let's look again at what has come to be a very provocative model system, one that has inspired much current cancer research. You may recall our earlier discussion of certain viruses that prey upon bacteria. They inject their DNA into the bacteria, and thereafter all the bacterial machinery is committed to producing more viruses.

Well, sometimes, after the virus's DNA enters the bacterial cell, something quite strange and unexpected happens. The DNA of the virus quietly tucks itself right into the DNA of the bacterium: the virus genes combine with the bacterial genes. And no new viruses are made. The bacterial cell goes on dividing as though nothing had happened to it. But something profoundly significant *has* happened. The infected bacterium and all its descendants contain the virus's DNA, and have altered properties and behavior as a result of it.

What's going on? Well, the genes of the virus, now a part of the bacterium's DNA, are functioning. They are dictating the making of messenger RNAs, and these messenger RNAs are going to the bacterium's ribosomes and dictating the construction of new proteins. These proteins become part of the bacterium, thereby altering its character. In sum, the bacterium and all its descendants have *changed* by virtue of the presence of virus genes, now a part of the bacterial DNA.

This is indeed ominous behavior on the part of viruses. Its significance for cancer is sharply heightened when we learn that many cancers — in animals — are caused by viruses. And cancer viruses in animal cells do things remarkably similar to what bacterial viruses do in bacterial cells. They enter cells, seem to disappear, their genes combine with the cell's DNA, and the cell's properties are permanently altered. The alteration in this case is to malignancy.

If we put the virus and mutation stories together we can make a significant generalization: *new* genes from viruses or *altered* genes from mutations may cause new proteins to be formed in cells. These in turn may trigger more rapid growth, surface changes leading to asocial behavior, and other cancerlike properties.

Many Cancers Don't Seem to Be Caused by Viruses

There are a large number of cancers, including all *human* cancers, that don't *seem* to be caused by viruses. This doesn't mean that human cancers aren't caused by viruses. It just means we aren't able to show it. Indeed, the phenomena we've described above would make it very difficult to find viruses in cancer: the virus sometimes seems to be quite clever at hiding its presence.

Body Response to Cancer

When I say cancer cells are unneighborly, asocial, and tend to dissociate from one another, you will appreciate that this is a *surface* phenomenon. Cancer cells "feel" neighboring cells by cell-cell contact, by surface-to-surface interaction. This means that the surfaces of the cancer cells must have become different from the surfaces of the cells from which they were derived. This has been shown by experiment to be so.

Now, if the cancer-cell surface differs from its normal counterpart, is it "foreign"? By this I mean: Is the surface different enough to appear foreign to the body's protective immune system? Well, the answer seems to be yes. Cancer cells do seem to provoke immune responses — that is, the body's defenses weakly react by trying to destroy the cancer cells. This information offers hope. For if the body defends itself against cancer, it may be possible to strengthen defenses with a vaccine, using principles familiar in the treatment of infectious diseases.

Cancer and Our Environment

A steadily accumulating body of evidence supports the view that cancers are caused by things that we eat, drink, breathe, or are otherwise exposed to. The evidence is of three kinds. First, the incidence of many types of cancer differs greatly from one geographic region of the world to another. Second, when groups of people permanently move from one country to another, the incidence of some types of cancer changes in their offspring. For example, when Japanese move to this country, the relatively high rate of occurrence of stomach cancer they experience in Japan falls so that their children experience such cancer only a fifth as frequently, the same incidence as in other Americans. Orientals have low incidence of breast cancer, but when they come to the United States, it increases sixfold. Third, we are becoming aware of an increasing number of chemical pollutants in air and water and food that have been proven to be cancer-producing.

This knowledge is encouraging in one sense, for it offers promise of our being able to eradicate cancer by controlling environmental pollution. But we know that these things are difficult to accomplish. Witness the complete failure to reduce cigarette smoking, in spite of full knowledge for fifteen years that it is the major cause of one of the most malignant of all cancers, that of the lung, which kills nearly 100,000 people a year.

Immediate Cause versus Distant Cause

It should be understood that virus, mutation, and environmental causation of cancer are not in conflict, or contradictory. Mutations and viruses are direct ways of changing DNA, thereby causing cancer. But mutations must be caused *by* something — and it may be that environmentally dispersed chemicals are the guilty agents, getting into our bodies and there altering DNA. And viruses may require activation by environmental chemicals before they can cause cancer. In this sense, the environmental chemical is a distant cause; the mutation or the virus, the more immediate cause.

It would be a great mistake to think that because environmental chemicals cause cancer, we can abandon our search for the immediate cause or causes. There is no assurance whatsoever that we can control our environment. And cancer plagued us long before industry poisoned our environment. The only sure way of ultimately realizing a prevention or cure is to learn in detail what is going on inside cells as they make the deadly transition to malignancy.

CHAPTER IX

The Search

The knowledge I've shared with you in this book has been discovered by our fellow humans mostly during a tiny portion of evolutionary time, the last few hundred years. If all of life on earth were compressed into one year of our calendar, man's effective knowledge-gathering period would be but a few seconds of that year. How did we humans manage to find out so much about ourselves in so short a time? And how does the process relate to other truth-seeking endeavors of humans? In this chapter we'll ponder these things.

The Knowledge-Gathering Process

The astonishing accumulation of knowledge for which we humans have been responsible has its roots first and foremost in our intrepid, inborn curiosity, our insatiable thirst for explanation. From earliest times, faced on all sides by mystery, we urgently tried to dispel it by understanding it. Whether it was managing a home, practicing war or agriculture, navigating upon the sea, exploring the land — in all human endeavors, the unknown was an unwelcome companion. The rewards for successful knowledge-gathering were the abolition of uncertainty, doubt, and fear; the comfort of comprehension; the pleasure of confident prediction; the satisfaction of the need to put things in order, to make neat arrangements; and the acquisition of power — power to control nature.

Observation

Curiosity compels action, action in the form of exploration. We're exploring when we strike stones together in search of a flame, push a dog sled through hundreds of miles of ice to reach the top or the bottom of the earth, walk miles straight up a mountain or dive miles straight down into the sea, construct a machine in which we can fly, painstakingly record the look and behavior of thousands of fruit flies in order to learn about inheritance, build models with wire and pieces of cardboard to work out the structure of DNA. Exploring means observing: watching, feeling, smelling, tasting, prodding, listening, questioning. And we write down the things we observe so as not to forget them, or lose them, and so that we may tell others about them — accurately.

Ideas

Now curiosity and fact-gathering weren't enough to make the unknown known. Other animals had curiosity but didn't build knowledge. Our peculiar brand of the search for explanation was coupled with an ability to conceive ideas, and to test them by experiment. An idea (or theory, or hypothesis) puts observations in a meaningful arrangement, envisions ways in which facts might relate to one another. An idea tries to make sense out of things that don't make sense. Think of a situation in which you've been presented with some information; say, on a newscast. Your mind quickly tries to clothe the facts in meaning, give them significance in a larger framework, envision a causal relationship between them. For example, if several similar events have occurred within a short period, like a series of oil spills from sinking or damaged tankers, you may try to ascribe to them a common cause. The mind abhors coincidence; it insists on seeing causality, on asking why.

Ideas are related in a complex way to our brain's remarkable capacity for imagining. Imagination allows us to picture interrelationships that might be. Once the picture has taken shape, we have a theoretical explanation for what we've observed — an untested idea.

Testing Ideas

Now, of course, an idea by itself, no matter how clever, is simply something you think up. It is a creation of your intellect — just as the earth produces a flower, or a weed, or for that matter as a fire produces smoke! It is not necessarily true; in this sense it starts its life soft and mushy. Ideas get hardened and tempered by experimental verification. For an idea to correspond to truth you must test it to find out if it is truly congruent with nature. The best way to know if an idea fits with the facts of nature is to see if it allows you to *predict* something. Not by luck, but consistently. If an idea states that certain things are causally related, then if one does something to the things, something foreseeable will happen. If that something doesn't happen, the idea is defective. A defective or wrong idea is not necessarily a bad idea. It may get you started doing a series of experiments that, though proving the idea wrong, leads you to a new, truer insight. The experiments with the plants I described in the last chapter, while starting off with a wrong idea, revealed the fascinating ability of beryllium partially to substitute for magnesium in plant growth.

Suppose your car's engine fails to function. After exploring under the hood you produce the idea that the failure is due to a faulty fuel pump. You predict that if you replace the pump, the engine will function. If your prediction is correct, your idea has become respectable — because it is a statement of truth. Your idea becomes even more respectable if it allows you to generalize beyond the particular instance: *any* failing car with a certain set of symptoms will require a fuel pump replacement.

The way you test your idea is to do an *experiment.* An idea is only as good as the experiment devised to test it. Good experiments can be simple, as in the case of replacing your car's fuel pump, or they may require much ingenuity and imagination. Of great importance is the choice of a good model system. For example, observations made on inheritance in humans produced many ideas in explanation, few of which could be tested directly. Model systems were needed that were simpler than humans, more manipulable than humans, and in which a generation was a few hours instead of thirty years. So it came about that peas, fruit flies, bread molds, and bacteria gave us the base of knowledge for understanding human genetics.

Repeating Experiments

A satisfactory prediction-confirming experiment is a profoundly gratifying experience — especially if you're the one who performed it. But the job is not yet done. Scientists must repeat their experiments in different ways to be sure there is no possibility of error. They will seek out colleagues and ask them to find flaws in their ideas and experiments; they will attend scientific meetings and present their findings to audiences of critical colleagues. And finally they will publish their work in sufficient detail so that the international scientific community can learn about and repeat the work. Most scientific discoveries become widely accepted as truth only after the experiments have been confirmed by several scientists. Only then does the meaning our ideas give to observations become truth.

None of this is easy. An important truth that scientists learn is how extremely difficult it is to prove conclusively that something is true.

Being Ready for the Unexpected

Having described to you what I think you'd agree is a thoroughly logical process, I must now tell you that unexpected things happen with disconcerting frequency. Indeed, they happen sufficiently often so that another essential attribute of the knowledge-gatherer must be *alertness for surprise.* Unexpected findings may be due to error in designing experiments, to the fact that the original idea was wrong, or they may have some trivial explanation. But above all surprise in science must be expected by the very nature of science's mission. The subject of science, after all, *is* the unknown.

A Tale of Discovery

You've been growing disease-causing bacteria on a jellylike material spread on the bottom of glass dishes. You enter your laboratory one morning and notice that one of the several dishes looks strange. They should all look like this:

where each little mound (dot) is a colony of bacteria that grew from a single cell. Instead, one of them looks like this:

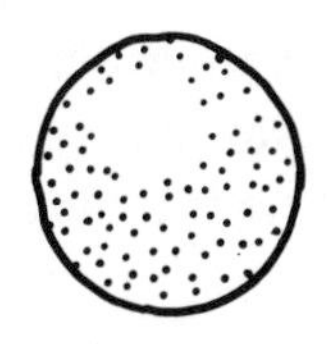

There's a big clear area where no bacteria have grown. The first impulse is to throw away the offending dish. But then this unexpected and irritating observation sets you pondering. You know bacteria don't discriminate against one area of the dish without reason. It seems that something prevented the bacteria from growing in the clear area. Could it be that something poisonous fell on the dish when you had the cover off yesterday? You seem to recall that the window had been open, and the room is a bit dusty. Some poison, brought to the dish by a bit of dust, could have diffused away in all directions to make the disk-shaped area of no bacterial growth. You decide to test this idea by getting samples of dust from around the room, spreading out some more bacteria, and putting a tiny bit of dust in the middle of the dish. Two days later the bacteria are growing fine.

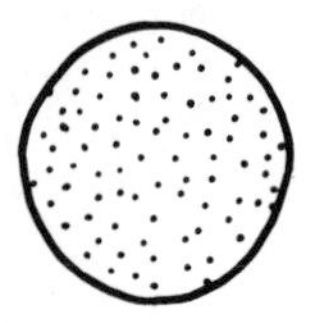

Back to the drawing board! You now notice that at the back of the bench lies an ancient, moldy peanut butter sandwich. Excited, you surmise that something from the sandwich got on the dish and poisoned the bacteria. You take a tiny portion of peanut butter and a tiny crumb of bread and test them as you did the dust. Two days later the dishes look like this:

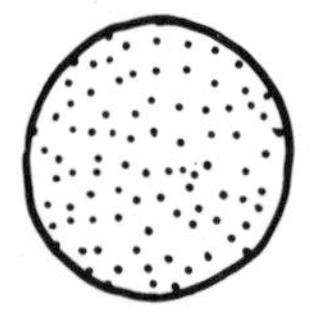

In disgust you grab the dishes and sandwich and throw them in the disposal bin. And decide to wash your hands of the whole miserable business. In doing so, you notice that one of your hands has a little bluish green stain on it, apparently from the bread mold on the peanut butter sandwich. And now you experience a prickly chill up your back into your scalp, and a wild idea brazenly leaps up. Could *bread mold* have got into the dish? Quickly you take a tiny scraping from the moldy crust of the sandwich and put it on a dish over some more bacteria. The next forty-eight hours move slowly, for you have a gut feeling something's going to happen. And it does.

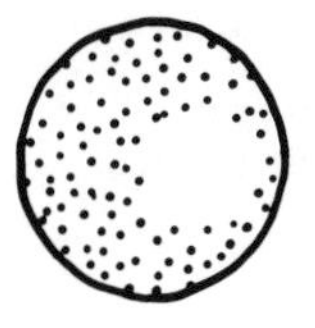

Bread mold won't allow these disease-producing bacteria to grow.

This little fiction illustrates some of the features of scientific discovery we've discussed. We'll end it here because the true story has already provided the best possible ending. Although Sir Alexander Fleming was not partial to peanut butter sandwiches, as far as I know, he did discover penicillin in 1929 in a manner similar to the one I described.

Avery's Famous Experiment Again

Let's go back to Avery's experiment, which proved that DNA was the material basis of inheritance. You'll recall, from Chapter II, that Avery had used a model system: pneumonia-causing bacteria. He had *observed* that a mixture of molecules released by dead pneumonia-causing bacteria turned living benign bacteria into pneumonia-causing bacteria. His *idea* was that the responsible molecules in the mixture were DNA. The first *experiment* he designed was one in which he exposed the mixture of molecules produced by the dead bacteria to an enzyme that breaks up DNA. If the active principle was DNA, he *predicted*, the enzyme treatment would destroy the material's ability to transform benign bacteria to pneumonia-causing bacteria. That's exactly what happened. That very simple experiment confirmed the prediction, and a new truth, destined to influence much future scientific experimentation, was disclosed. From that point on, though there remained much to do to clinch the proof, there was little room for doubt that DNA was the material basis of inheritance.

Limitations of Science

Ideas, we have said, are often predictions that suggest experiments by which they can be tested. When one can regularly predict how certain events are going to come out, the ideas upon which the predictions are based become widely accepted as truth, and are called *laws* or *principles of nature.* These are the truths we have discussed in this book. On the other hand, what we tend to call "good" ideas in our everyday life, and in sociology, psychology, and in philosophy, art, and religion — ones we judge to be inspired, brilliant, or ingenious — are not necessarily good by the criteria of science. For most such ideas are generated in response to phenomena that are too complicated to be tested by experiment. Predictions from ideas of this kind seldom come out the way the predictor hopes, except by luck.

When we say that a great painter has revealed *truth* through a painting, that psychology and psychiatry have disclosed truths about human behavior, that the theologian has discovered the truth of God's existence, we use the word "truth" in a way different from science's use of the term. None of the criteria we've discussed are applied in such situations. We might more accurately say that these seekers have produced explanations that give many people an intuitive feeling that a truth has been revealed. But as many other people remain unconvinced. Universal acceptance of such truths is not possible.

Indeed in the complex, mysterious realm of human conduct and values, ideas are "a dime a dozen." They are generated freely by all of us, eagerly trying to make sense out of the bewildering display of human thought and action we see about us. Ideas appear in such profusion partly because they are unchallengeable, untestable. They attract some because they cannot be proved wrong; they repel others because they cannot be proved right. The tough challenge, throughout history — in all fields of human thought and exploration — has been to produce ideas that can be tested.

The human dilemma is in part based on the fact that personal or social actions *must* often be taken under the limp banner of untested and untestable ideas. Persons and governments have to make decisions on matters about which knowledge is very limited. Be that as it may, the extent to which political decisions can be arrived at rationally — that is, taking into account existing available knowledge — is a measure of wisdom. Rationalism may be said to be the application of the principles and methods of science to problems bigger than those science would usually choose to tackle.

I don't want to imply that science is more virtuous than the more complex fields concerned with human behavior. Science simply asks simpler questions; it deliberately limits its search to the asking of *small* questions. These lead to clear, limited answers. Many small questions generate many small answers: each one validatable by any and all skeptics willing to go to the trouble of repeating the key experiments. Science, then, sets its own limits: the truths it proclaims must be verifiable by experiment or they are not worth proclaiming. The long-term impact of this kind of truth-harvesting on the lives of all humans has been tremendous because it is cumulative and enduring.

Mental Illness and the Chemical Brain

Humans have, since earliest times, been fascinated by the workings of the human mind. Mental illness, in particular, has both frightened and intrigued us, and urgently demanded explanation. Some past ideas have attributed mental illness to gods, devils, complex social and family interactions. As we have already noted, these explanations can seldom be tested by experiment because most aspects of the workings of human consciousness are too complex to be dealt with by scientific methods at the present time. In spite of this, many such ideas are erroneously widely accepted as truth and as the basis of psychiatric treatment.

The history of man's concern with mental illness has some ironies and unexpected twists. It illustrates what we said earlier about the appearance of the unexpected. As man has struggled down the years, with the expected lack of success, to understand and treat mental illness on its own terms, that is, by various forms of psychotherapy, an increasing body of scientific evidence has accumulated, showing that human behavior is dramatically altered by chemicals. Consistent with all other evidence that life processes are chemical processes, an ever-expanding number of natural and synthetic chemicals are producing dramatic alleviation of symptoms of mental illness. Even the distressing, widespread *misuse* of chemicals in the form of drugs in our society attests further to the chemical nature of mental processes.

Years ago it was discovered that a baffling psychosis called pellagra could be completely and permanently relieved by administration of vitamin A. Overnight, research transformed a mysterious mental disease into a vitamin deficiency. Another severe and very common schizophrenia-like psychosis was found by researchers to be treatable with an antibiotic. The disease was syphilis.

It was discovered about twenty years ago that a simple salt, lithium carbonate, when taken regularly by mouth, prevented the appearance of the symptoms of manic-depressive psychosis. Soon the symptoms of this dreadful and all-too-common affliction suddenly became treatable. The effect of lithium on mania and depression was an empirical observation, not predicted from knowledge of brain chemistry. It is interesting to note, however, that lithium is a very close relative of sodium, and sodium had long been known by scientists to be essential to brain function. We do not yet know the mode of lithium's action.

So it is that largely fortuitous discoveries of the effects of chemicals have opened up human behavior to critical study by scientists. A dramatic upswing in the relief of distressing symptoms of mental illness has been the result, and we may expect this kind of progress to continue.

Basic Research and Applied Research

Throughout this book I have used the words "science" and "research" to mean *basic* research: exploration for new knowledge. There is a much larger domain of research endeavor called *applied* research, or technology, which has to do with the application, to human needs, of knowledge gained from basic research.

Basic and applied research are done quite differently. In applied research, since the basic knowledge base is available, specific production goals may be set, teams of researchers may be assigned definite tasks, leaders may easily assess performance, contracts may be let, and business can risk capital in the venture. Thus it was, for example, in planning and accomplishing flights to the moon and Mars, or producing massive quantities of polio vaccine to immunize the public.

Basic research, in contrast, is the exploration of the unknown. Guides are not available; the researcher is alone with his ingenuity, imagination, and curiosity. Unexpected occurrences are the norm, and the researcher must be prepared to exploit them. There can be no timetables. Performance can be evaluated only after several years, during which the scientist must freely develop skills and hypotheses.

Applied researchers must use their inventiveness to make known principles perform specific jobs for them. Basic researchers must discover the principles.

It hardly needs saying that technology would be a

helpless, bumbling giant if there were not a steadily advancing front of new knowledge. Look at medicine today to see the nature of the problem. Basic research has made possible a remarkable series of major advances in the elimination and treatment of many diseases. But the diseases that remain to harass us represent vast domains of the unknown. Nothing but new knowledge will halt the onslaughts of cancer, heart disease, stroke, genetic disease, and other ills. Yet government and private applied research exuberantly produces more and more elaborate and costly machines, inevitably benefiting a very few patients at steadily increasing cost to all of us. Thus, surgery for cancer, kidney dialysis machines, artificial hearts, and a variety of other ingenious applications of known principles are costly stop-gap measures that dramatize the deficiency of our present knowledge of disease.

The cost of medical and health care to Americans is now $150 billion a year and is steadily rising. This appalling bill is the price we pay for an excess of applied research over basic research (as well as for some other things like bad planning, inefficiency, and greed). Under these circumstances, you would surely expect America to be making a healthy investment in basic research so as to be able to prevent and cure disease. Not so! About one half of 1 percent of our health bill goes to basic research, and our nation's brilliantly promising research establishment is slowly being strangled for lack of funding.

Supporting Science

It wasn't always this way. Thanks to an enlightened policy of our government in the 1950s and early 1960s, we learned how to discover good basic scientists and support them. The present system is an outgrowth of the earlier generously funded one. Traineeships and fellowships, to the extent that funds are available, provide salaries for highly promising students, both before they obtain their graduate degrees and for a few years after. This allows a student to work in a laboratory under a preceptor of his choice until his training is complete and his area of specialization is delineated.

The student may then apply for a grant to support him as an independent investigator in a university or research institute. This grant application is the young scientist's definitive statement concerning the problem he wants to tackle, his ideas, projected experiments, and his belief that his work is important to man's health. Such an application is a major work for a scientist — he puts his future on the line, mustering all his imagination and skill to the task.

These applications are carefully reviewed by a group of scientists who serve the government as consultants. (The process is known as peer review.) These scientists assign the application a priority rating on the basis of scientific merit. Grants are then awarded until the available funds, authorized by the Congress, are used up. Once an award is made, the researcher exercises considerable autonomy in carrying out the work and publishing his results.

This system of supporting basic science is the finest that has ever been devised anywhere. It ensures selection of the most promising projects, it encourages independence and originality, puts a high premium on self-motivation and originality, holds the scientist accountable, and yet allows reasonable scope for flexibility and the pursuit of the unexpected, essential in all exploratory endeavors.

Unfortunately, as I have said, for some years this great science-support system has been under attack from members of the executive branch and of Congress who see the support mechanism as "unbusinesslike" and the peer-review process as "cronyism" (because scientists are giving scientists money!) and who see science as taking too long to solve the medical problems it has been "charged" to solve! Some particularly unjust attacks come from congressmen who lack understanding of the importance of model systems, and see some research proposals as "irrelevant." We may suppose that such critics would have been equally outraged if Mendel had submitted a proposal to study peas, Morgan to study fruit flies, and Avery to study bacteria!

This neglect of the nurture of medical science and the attack on its methods in recent years has resulted in the present pressed state of biomedical research support. The training of young scientists languishes and the great project grant system is slowly being enfeebled. At the time of this writing, about one third of all grants are being funded. The other two-thirds represent the crushed hopes of a large number of brilliant young American medical scientists. The search for biological knowledge, the only hope for relief from the suffering caused by disease and from the outrageous cost of medical care, is being blocked by human irrationality.

The Uses of Knowledge

Knowledge, and the methods by which it is gained — science — are morally neutral. Nature's secrets are there, and men and women will seek them out. But the way society uses knowledge is seldom morally neutral; for knowledge is power, and the desire for power, to do both good and ill, has gripped humanity from the beginning.

In the health area alone, our thirst for knowledge has given us a dramatic increase in life expectancy; the opportunity for women to control their childbearing; the virtual elimination of such diseases as plague, cholera, tuberculosis, polio, smallpox, diphtheria; vitamins and general improvement in nutrition; x-rays in medical diagnosis and treatment; extensive improvements in surgical and anesthetic technology; advances in the prevention or treatment of numerous immunologic, hormonal, neurologic, and genetic disorders. This is indeed a monumental achievement for medical science. And because of recent further advances in basic science, notably the important discoveries described in this book, we have every reason to be optimistic that we will be able to eradicate or treat more effectively other major causes of human misery, such as cancer, genetic disease, and cardiovascular disease (if national policy permits).

On the darker side, we threaten the very structure of the world's pool of DNA with radiation from proliferating nuclear devices; we poison and pollute our water, food, and air with an alarmingly increasing number of industrially profitable chemicals; we destroy the ozone layer, which protects all creatures from deadly radiations from the sun; we spawn more babies than the earth can feed though birth control methods are available to all. Humans seem to be in a race with themselves to see whether they can achieve ultimate bliss or ultimate agony.

Should Science Be Controlled?

Science is simply the shedding of light on mystery; clarifying that which is already there in nature. But the emergent knowledge gives power to people and governments, and it will be used for good or evil according to society's values. If in a free society we desire to discourage evil uses of knowledge and encourage good uses, we must seek the solution in the values of people. This seems plain to me. But there are some who suggest that a simpler way is to suppress science. This can readily be done by withholding funds because modern research is so expensive to conduct. But try to restrict further the funding of biomedical research and observe the result. Humans will not be deterred from trying to understand the diseases that affect them. If the search for knowledge is suppressed we can turn only to applied research, which certainly is safer but much more expensive. For example, if we had been prevented from doing the basic research that led to the polio vaccine, we would have continued to try to find new and better treatments for victims of polio. The result of such abandonment of the search for new knowledge and forced reliance on contemporary knowledge is entirely predictable; medicine would be totally stalemated by being reduced to gadgetry, costs would continue to rise, and the ill would languish.

We are now, as never before in history, on the threshold of the application of the past forty-five years' progress to contemporary problems of human illness. The potential for good — further rapid advance in alleviating human suffering — is immense.

Benefit versus Risk in Medical Research

But what if scientists themselves urge the shackling of scientific exploration? An example of this is currently receiving much national attention. I am speaking of recombinant DNA research, which we discussed in Chapter V. There we learned that it is possible to link pieces of DNA from any animal or plant to the DNA of bacteria, after which the bacteria multiply and produce many copies of the added DNA. The bacteria simply serve as factories to make lots of copies of any chosen piece of added DNA. Biologists consider the technique one of the most valuable ever devised for expanding our understanding of genes: their structure, and the manner by which they are turned on and off. Having explored with me the ripe problems of embryogenesis and cancer in the two previous chapters, I hope you began to imagine the value of this new technique for exploring the great problem of gene expression.

But something appears to be amiss. For there are a few vocal scientists who see hazards in doing these experiments. Could the added genes alter the bacteria to a form dangerous to humans? Could such experiments unnaturally change evolution by creating novel, potentially dangerous forms of life?

Well, it is certainly a theoretical possibility that a more dangerous bacterium could be created if DNA were added to it. Indeed, we've learned in this book that cells can be changed by added DNA. The fact that bacteria have never been made more dangerous by the addition of DNA doesn't prove that it couldn't happen. For these reasons, scientists have, on their own initiative, agreed not to reproduce cancer virus genes and genes from dangerous bacteria in other bacteria by recombinant DNA techniques. In addition, other types of experiments that involve multiplying genes from humans, or from animals closely related to humans, in bacteria are to be done with special precautions.

When people agree to take precautions, it is assumed

that there are real risks; that is, risks that can be anticipated on the basis of previous occurrences. But in the case of recombinant DNA, the precautions I have noted have been taken against hypothetical risks – risks that are only imagined. What have we learned in these pages that can help us to evaluate these conjectural chances of producing a new organism that could cause us harm? Well, I feel safe in saying that the essence of what we've learned is that, though anything is possible, such an occurrence is extremely improbable. You'll recall we learned that almost always a change in an organism's DNA is detrimental to it; that is, it leads to a reduced capacity to survive. By way of analogy, random additions of sentences to the plays of Shakespeare are not likely to improve them! This statement does not contradict the corollary fact that the evolution of an organism *requires* changes in DNA; that is, that alteration of, or additions to, DNA produce improvements. But, as we learned in Chapter V, these events are rare. The principle that DNA changes are harmful by virtue of reducing survival chances applies whether a change in DNA is caused by a mutation or by some foreign genes we deliberately add to it. So, added foreign DNA, rather than making an organism more virulent (dangerous), will do quite the reverse; it will almost invariably make the organism weaker.

But in addition to the general observation that altering a bacterium's DNA almost always leads to an organism less fit for survival, there's another important consideration that reduces the risk even further. Evolution and genetics tell us that the construction of a disease-producing organism is an extremely complicated affair. Bacteria like those that cause typhoid, plague, diphtheria, tuberculosis, and so on are complex organizations of genes, annealed in the crucible of billions of years of evolution. To presume that we humans, in a few blundering years of juggling genes, could create similar or better gene combinations is colossally presumptuous.

As I hope was made clear in earlier chapters of this book, mutations and the mixing of DNA are random, and so evolution has been determined by a series of chance events. There is no biological justification for assuming that human manipulation of DNA has special significance relative to nature's own manipulation of DNA over the last 3 billion years. Nor is laboratory mixing of DNA in living organisms new. Since the 1930s we've been adding DNA to bacteria and causing inherited transformation, as in Avery's experiments. Recombinant DNA experiments have gone on without precautions for the last five years. And we have reasons to believe that gene-mixing happens frequently in nature.

The benefit versus risk element in recombinant DNA research — exploration of the unknown with an immensely powerful tool for the advancement of knowledge versus conjectural hazards to man — is just one of many kinds of similar challenges we must deal with in our complex society.

We humans have of course faced benefit versus risk challenges since the dawn of consciousness. The primitive hunter had to weigh the benefit of killing his prey, and so obtaining a meal, versus the chances of being killed himself in the process. Now, in our private lives and in the government decision-making process such choices are much more complicated. Generally, the benefits are far more dubious than recombinant DNA research, and the risks are *real*, not imaginary. For example:

1. Industry's exuberant outpouring of consumer goods,

versus the desecration of our water, air, food, and even our ozone layer.

2. Industry's production of "drugs" — chemicals supposed to make people feel better — versus the risks inherent in consuming quantities of inadequately tested, often useless materials.

3. The production of vaccines, such as that for polio, versus the clear risk of untoward effects in the recipients in the future. After all, no drug or vaccine can be checked for safety for thirty years before being put on the market. Yet we know that some of the most serious untoward effects of medications may occur only after long latent periods.

4. The development of nuclear power versus the clear and present danger to our DNA of exposure to radiation.

5. Federal subsidization of the tobacco industry versus the certain knowledge that its products are killing 100,000 Americans a year from cancer and untold additional numbers from other ills.

These are, of course, only a sampling of society's choices. You can conjure up many more examples from the experiences of your own life. But in general there is little we do that doesn't involve some risk.

Clearly it is incumbent upon all citizens to weigh as best they can benefits versus risks so as to be able to make informed choices. A felicitous example of this occurred in Cambridge, Massachusetts, in 1977, in connection with recombinant DNA research. The City Council, alarmed by extreme statements about the danger of recombinant DNA research in prospect at Harvard University, asked a citizen's committee to hear expert testimony, study the problem, and recommend action. The committee, composed entirely of laymen, acting with diligence and responsibility, recommended that the research be continued with the observation of certain reasonable safeguards in addition to those required by the NIH. The City Council accepted the recommendation and the committee won the admiration of scientists and citizens alike. This happy outcome supports my conviction that laymen can comprehend important scientific subjects and issues and make responsible decisions relative to them.

The recombinant DNA story raises these questions: If we ever decide to proscribe the quest for knowledge because of perceived risks, do we then search only for knowledge that is without risk? What kind of knowledge is that? In exploring the unknown, how do we know what will be dangerous; what will be safe? Surely, the safe explorer is the one who stays in bed, no matter what the domain of his search.

The Future

Humans have created the terms of their own future evolution. Unlike any other living species, we alone profoundly modify our own environment — all too often to our detriment. Our fate is now determined more by what we've done to the world than by what the natural environment can do to us. We call this phenomenon *cultural evolution.* It's a completely new ball game. We can alter our thinking by drugs, poison our air and water and food, damage our genes by nuclear radiation or by ultraviolet radiation entering through an ozone layer we have destroyed, eliminate a variety of other animal species from the earth for all time, exhaust our energy supplies in producing things we don't really need. We can also prolong life, eliminate disease, alleviate poverty, make beauty, comfort, laughter, satisfaction.

The air we fill with poison we also fill with music!

Indeed, we have an almost infinite capacity both to create beauty and joy and to make monumental misery. The answer to whether we shall have the vision and will to make life better for all evolution's creatures lies hidden in the shrouded future. But of one thing we can be certain: the society that stifles the free exercise of curiosity has little to offer the future. The search for explanation is as basic a drive in us as is hunger and sex. We must continue the search, and in the search itself find our reward.

Our knowledge is like a great living library — knowledge gained through the ages is there for all to examine, and new volumes continue to find a place upon the shelves. Between the covers of the books is everything we indisputably know about our world. It is the foundation for the building of future knowledge. We may wish we had more volumes to help us understand and so control human greed, for example, or to acquire love and wisdom, but they've not yet been written. The absence of the unwritten volumes shouldn't diminish the value of the ones on the shelves.

There are some who say that knowledge gained through science "dehumanizes" life by shining a harshly dispassionate light on the unknown, on mystery. To me the contrary is the case. By knowing what science has revealed we cannot but expand our sense of wonder at the incredible beauty and "ingenuity" in the design of us. By being able to look beneath our skins to the busy commerce between DNA and RNA and protein, have we lost or gained? I hope my readers will feel that they have gained. And for you who take no joy in the machinations of molecules, and who need the mystery of unexplored nature to nurture your personal "truths," be comforted to know that science has only lightly scratched the surface of the unknown. The knowledge that still remains to be discovered vastly exceeds the knowledge that has thus far been revealed. There is as much room as ever before for wonder, beauty, inspiration, dreaming, magic, mystery, and the gods of your choice.

Terms Used in This Book

Index

Terms Used in This Book

Chapter I

Atoms: the smallest entities of which living material is made. There are over a hundred different kinds of atoms, but the five principal ones are carbon, hydrogen, oxygen, nitrogen, and phosphorus.

Molecules: atoms linked together chemically. Their average size is about 10 times larger than that of atoms.

Nucleotides: molecules making up the links in DNA and RNA chains. In DNA there are four: adenylic acid, guanylic acid, cytidylic acid, and thymidylic acid. In RNA they are similar, except that thymidylic acid is replaced by uridylic acid.

Amino acids: molecules that are the links in protein chains. There are twenty of them. They are often designated by the first three letters of their names.

Entropy: a chemical term for the relative state of disorder of a system.

Energy: a chemical term for the capacity of a system to perform work.

Chapter II

Information: an arrangement of symbols that instructs a machine to construct something.

Gene: a piece of information instructing the cell's machinery how to make a particular protein. Groups of genes instruct in the making of groups of proteins, which determine inherited traits.

Genetics: the science of inheritance.

DNA: a long chain of nucleotides. It is the chemical form of biological information and the substance of genes.

Proteins: chains of amino acids in specifically arranged order. Most of life's structure and function is proteins.

RNA: a chain of nucleotides similar to DNA.

messenger RNA: an RNA copy of one gene-length of DNA.

Ribosome: a combination of RNA and protein that, with the help of transfer RNA and a supply of amino acids, "reads" messenger RNA, linking the amino acids together in proper order to make a protein.

transfer RNA: many small RNA molecules to which amino acids become attached before being conveyed to the ribosome to be connected.

Bacterium (plural: bacteria): a one-celled form of life, much smaller and simpler than animal cells. Often able to live on nothing but simple salts and a sugar as a source of energy.

Virus: a combination of DNA (sometimes RNA) and

protein that can reproduce itself only inside a living cell.

Chapter III

Ozone: three oxygen atoms bound together. These molecules accumulate above the earth's atmosphere and form a protective shield against ultraviolet light.

Enzyme: a protein molecule that can do a specific chemical task. Acting as catalysts, enzymes make reactions go faster.

Membrane: a combination of fat and proteins that envelops the contents of a cell and protects it from the environment.

Chapter IV

Chlorophyll: the green-colored molecules of plants. They trap light energy.

Chloroplasts: compartments inside plant cells where trapped sunlight is converted to ATP.

Mitochondria: compartments inside cells where sugar molecules are burned to make ATP.

ATP: adenosine triphosphate. The form of useful chemical energy in cells that makes all the cells' work possible.

AMP: adenosine monophosphate: ATP minus pyrophosphate.

PP: pyrophosphate: two phosphates attached to each other. PP plus AMP is ATP.

Combustion: combining a molecule, such as a sugar molecule, with oxygen to release energy (heat or work).

Electron: a negatively charged part of an atom that, when moving, is electricity.

Chapter V

Evolution: the process by which present forms of life developed from the earliest form of life.

Mutation: alteration of the structure of DNA by a physical or a chemical agent. A mutagen is an agent that can cause mutation. Mutagenesis is the production of mutation.

Plasmid: a small, circular piece of DNA, carried by a bacterium, that can go in and out of the bacterial cell.

Recombinant DNA: an end-to-end combination of two DNA chains from different sources. More specifically, a piece of DNA spliced to a bacterial plasmid.

Chapter VI

Selection: the process by which the environment favors or disfavors a particular variant of an organism.

Chapter VII

Embryo: an organism in the early stages of its development.

Gene expression: the manifestation of a gene's being translated into protein.

Repression: the shutting off of a gene so that it can't be translated into protein.

Repressor: a protein molecule that prevents a gene from being expressed.

Bacteriophage: a virus that uses bacteria to make more of itself.

Regeneration: restoration of a removed organ.

Chapter VIII

Carcinogenic: cancer-producing.

Cancer virus: a virus capable of converting a normal cell to a cancer cell.

Index